AF657307

Engineering Economics and Costing

Second Edition

Engineering Economics and Costing

Second Edition

CMA Anil Kumar Mukhopadhyaya

Certified Value Specialist–Life

TECHSAR™

NEW DELHI

Published and Printed by
TechSar Pvt. Ltd.
4435-36/7, Ansari Road, Daryaganj
New Delhi-110002 (India)
E-mail: tech_info@techsarworld.com
Website: www.techsarworld.com

ISBN: 978-93-90620-15-9

© 2021 I.K. International Pvt. Ltd.

Reprint 2025

All rights reserved. No part of this publication may be reproduced, stored in a retrieval system, or transmitted in any form or any means: electronic, mechanical, photocopying, recording, or otherwise, without the prior written permission from the publisher.

Dedication

My Sons & Daughters-in law

Dr. Ashis Mukhopadhya & Dr. Indrani Mukhopadhya

&

Atish Mukhopadhyaya & Mitali Mukhopadhyaya

TATA

Date : 16-3-2021

FOREWORD

I have known Mr. Anil Kumar Mukhopadhyaya for the past twenty eight years, during my tenure in Tata Motors, Jamshedpur. He was in the Productivity Services department and was closely associated with various cost-optimization initiatives like Suggestion Scheme, Small Group Activity (SGA}, Key Result Areas (KRA} for the Axles Division, where he was also directly associated as facilitator of various Value Engineering groups.

I, in my various roles, have continuously sought guidance of Mr. Mukhopadhyaya on various Direct Material Cost Reduction initiatives with a focus on Value Engineering.

I have been keenly following a range of articles that he has written in connection with various cost-effective techniques in various national and International Journals. He already has 6 books on Value Engineering to his credit which is a huge contribution to society and Value Engineering.

The present book titled, **"Engineering Economics and Cost"** will be a valuable resource in the hands of not only the students community but also for all Value Practitioners who are actively seeking to manufacture cost effective products and services.

I wish him all the success in his endeavor.

Vishal Badshah,
Head - Jamshedpur Plant,
Tata Motors Limited.

TATA MOTORS LIMITED
Jamshedpur 831 010
Tel 91 657 228 1600 Fax 91 657 228 1600 www.tatamotors.com CIN L28920MH1945PLC004520
Registered Office: Bombay House 24, Homi Mody Street, Mumbai-400 001

Preface

During my working life I had the opportunity to take classes for the management people who are mostly technical on different initiatives, like Quality Circle, Suggestion Schemes, Kiazen, and Value Engineering which are useful for cost effectiveness. I observed that our technical associates are highly skilled in technical matters but lack knowledge regarding the cost aspects. In these connections I want to express one experience which speaks about the environment prevailing that time.

One of my friends was working in the finance in the year 1974. I asked him about the cost of one of the product manufactured by the company. He told me that cost of the product was a secret element and he could not share it with me. He advised me that if I really wanted it then I should go through the proper channel.

After two or three years whole environment was changed. Now the environment was so congenial that even the worker should also know the cost of his operation because he is responsible for generating the cost of the product. This has forced me to write this book which includes various aspects of cost engineering also.

After my retirement I took up my consultancy and I had a chance to meet different types of industries. I observed that the managements are now interested to recruit engineers who are either knowledgeable about various cost-effective initiatives and also have awareness regarding the various cost techniques which will be required to improve the cost. The present book aims to fill up this void.

I have included in this book some techniques like "Design to Cost' and "Target Costing" which though old but now the essential considering. In addition to that I have also included a new technique like "Should Cost Analysis" which is gaining importance because the world has become a universal market.

CMA Anil Kumar Mukhopadhyaya

Acknowledgements

I still remember that in my college days, there was one subject named Engineering Economics in the final year. In my mind, I used to express my displeasure towards the board members who had decided to include this subject in the curriculum. This was because my thought was that for an engineer what is the use of Engineering Economics?

However, when I started my corporate life, I found how I was enormously wrong by not concentrating on Engineering Economics. In the corporate world, economics has a far higher prominence in comparison to technology. When I realized this aspect, I enrolled for a Cost Accountancy course and set out on a certification path. In this journey one of my friends, Mr. R.N. Patnaik supported me by joining along with me in the correspondence course conducted by the Institute of Cost and Works Accountants of India. I must acknowledge him now at this juncture for this valuable collaboration at the mid-stage of my career.

During 1980s, the company I worked in, which was a commercial vehicle manufacturing company, started thinking about cost effectiveness of the product. Though at that time there was no competition for our product. But it was the futuristic eyesight of the management to critically think about the product the company was manufacturing. The company started working on the cost effectiveness of the products. This is the time I must acknowledge Mr. P.N. Pandey, Divisional Manager, who gave me the responsibility to monitor the cost of different items on a monthly basis for our axles division. I also acknowledge Mr. R.N. Mishra who became the next Divisional Manager and who also had deep faith in my work.

At the end of the twentieth century when India became more globally oriented, my company resorted to "Value Improvement" initiatives. I got associated with different initiatives like Quality Circle, Kaizen and Value Engineering. While Quality Circle and Kaizen are mostly on the process improvement side, Value Engineering was used for Value Improvement of our products, which had two components, i.e., function and cost.

Some years later, I started my freelance consultancy in the first decade of the 21st century. During that time, I had a chance to interact and advise various organizations like auto manufacturing, IT sectors having a vertical called cost reduction, electronics industry and civil construction. Similarly, I had interactions with different people from a range of engineering disciplines, e.g., Mechanical, Electrical, Electronics and Civil Engineering. And in all the cases, I observed that all the professionals were very good in technology but very poor in the knowledge of cost. But the knowledge of cost was and is still the highest priority topic to be successful in the corporate world.

With the above-mentioned background, I decided to write this book. And I must acknowledge Mr. Surya Prakash Prabhaker, the young Certified Value Specialist who was most helpful in my endeavor. I must mention about my younger son Atish Mukhopadhyaya who is great support in all my writing pursuits. Dr. Ashis Mukhopadhya, who himself is a writer gave me a lot of suggestions to make it reader friendly.

My better half, Santa Mukhopadhyaya without her support I could not think to have progress in fulfilling this project.

I request all my readers to suggest me further requirements which will help to improve the book. The readers may contact me on: anilm13@rediffmail.com.

CMA Anil Kumar Mukhopadhyaya

Contents

PART I

Engineering Economics:

Concepts, Methods and Techniques

1 Types of Costing

In the business parlor, the following types of costing are used:

(a) Absorption costing

(b) Marginal costing

1.1 ABSORPTION COSTING

In this method the cost is collected when it has already been incurred. This is the conventional method of ascertaining the cost. This method is used for mainly reporting to all stakeholders. The cost details are put in a certain style. Table 1.1 indicates the details.

Table 1.1 Absorption costing

Particulars	*Total (INR)*	*Products*		
		A (INR)	*B (INR)*	*C (INR)*
Direct material cost	50,000	20,000	15,000	15,000
Direct wage	10,000	5,000	3,000	2,000
Direct expenses	5,000	2,000	2,000	1,000
Prime cost	65,000	27,000	20,000	18,000
Factory overheads	20,000	8,000	6,000	6,000
Factory cost	85,000	35,000	26,000	24,000
Office and administrative overheads	6,000	2,400	1,800	1,800
Selling & distribution overheads	4,000	1,600	1,200	1,200
Total cost	95,000	39,000	29,000	27,000
Sales value	1,00,000	35,000	30,000	35,000
Profit(+)/Loss(–)	5,000	(4,000)	1,000	8,000

The management can take various decisions based on the above information. The decision may be like:

1. Since product 'C is the maximum profitable, the company should use the full capacity for producing it, if the market demands.
2. The product 'A' is the original product of the company and was very popular in the market. Since it is an old product, the design is also old. Instead of discarding this product, the management can think of re-engineering it using the latest technology, processes and materials.

3. Thirty-two percent is the overhead of the products all together. The management can engage outside consultant to look into the overhead aspects of the company.
4. Similarly, there may be various options the management can think of.

1.1.1 Test on Chapter Two (Absorption Costing)

1.1.1.1 State Which of the Following Statements are True or False

1. Factory cost is not included in the absorption costing.
2. Profit/loss is visible in the statement.
3. Selling and distribution cost is an element of this technique.
4. Factory cost is the same as factory overhead.
5. Sales value is an important item of the statement.

1.1.1.2 What's Your Choice?

1. Take out the element of prime cost.
   (a) Direct wage
   (b) Factory cost
   (c) Selling overhead
2. Profit/loss is calculated by
   (a) Deducting prime cost from sales value
   (b) Selling and distribution cost should be added to factory cost
   (c) Sales value minus total cost
3. To get total cost, you have to
   (a) Add selling and distribution overhead with factory cost
   (b) Add office and administrative overhead and selling and distribution overhead and factory cost
   (c) Add office and administrative overhead and factory cost

Answers

True or false

1. (F) 2. (T) 3. (T) 4. (F) 5. (T)

What's your choice?

1. (a) 2. (c) 3. (b)

1.2 MARGINAL COSTING

The marginal costing is a very useful technique in the hands of the management. A number of decisions depending upon the situation can be taken with the help of this technique. Marginal costing system is also known as "out-of-pocket" costing. While in UK, it is popularly known as marginal costing but in USA it is known as direct costing.

In this technique the same cost structure is put into different fashions. In this technique, the materials and labor, which are variable costs, the fixed cost is also divided into variable and fixed costs. Variable parts of the fixed cost are added to other two variable costs. The following assumptions are made for the marginal costing.

Assumptions for marginal costing:

1. The fixed cost and the variable cost should not be changed during the period under review.
2. The product mix should remain fixed.
3. The sale value of the product should remain same during the period under review.
4. There should not be any inventory during the period.

The concept of marginal costing in the simplest form can be put as in Table 1.2.

Table 1.2 Concept of marginal costing

Elements of cost	*Cost of 100 Pcs (INR)*	*Cost of 101 Pcs (INR)*	*Increase of cost (INR)*
Direct material	50.00	50.50	0.50
Direct labor	30.00	30.30	0.30
Variable overhead	20.00	20.20	0.20
Fixed overhead	15.00	15.00	Nil
Total	**115.00**	**116.00**	**1.00**

The cost structure which has been used for the absorption costing is used in the form of marginal costing as shown in Table 1.3.

Table 1.3 Marginal costing

Particulars	*Total (INR)*	*Product A (INR)*	*Product B (INR)*	*Product C (INR)*
Direct material cost	50,000	20,000	15,000	15,000
Direct labor cost	10,000	5,000	3,000	2,000
Direct expenses	5,000	2,000	2,000	1,000
Prime Cost (I)	65,000	27,000	20,000	18,000
Variable Overhead				
Production	10,000	4,000	3,000	3,000
Administration	1,000	400	300	300
Selling and distribution	2,000	800	600	600
Total Variable Cost (II)	13,000	5,200	3,900	3,900
Marginal Cost (III = I + II)	78,000	32,200	23,900	21,900

Sale Value (IV)	1,00,000	35,000	30,000	35,000
Marginal Contribution V (IV – III)	22,000	2,800	6,100	13,100
Less Fixed Overhead				
Production	10,000	4,000	3,000	3,000
Administration	5,000	2,000	1,500	1,500
Selling and distribution	2,000	800	600	600
Total Fixed Cost VI	17,000	6,800	5,100	5,100
Profit (+)/Loss (–) [V – VI]	5,000	(–) 4,000	1,000	8,000

1.2.1 Mathematical Formula of Marginal Costing

Sales value = No. of sales unit × Sale value per unit

$$= N_s \times S_p \quad (1)$$

Marginal cost = No. of sales unit × Marginal cost per unit

$$= N_s \times MC_p \quad (2)$$

Marginal contribution = Sales value – Marginal cost

$$= N_s \times S_p - N_s \times MC_p$$

$$= N_s (S_p - MC_p) \quad (3)$$

$$\text{Fixed overhead} = F \quad (4)$$

Profit/loss = Marginal contribution – Fixed overhead

$$= N_s (S_p - MC_p) - F \quad (5)$$

1.2.2 How to Use Marginal Costing for Decision Making

Marginal costing technique can be utilized in various situations. Each situation will be dealt with proper examples.

1.2.2.1 Product Mix

Marginal contribution can be utilized to come out with a product mix with a certain constraint, may be like scarce raw material.

Illustration 1: M/S XYZ is a manufacturing company producing three products, Alpha, Beta and Gamma. All the three products require the same raw material. The yearly material availability is only 1,50,000 kg. The details about the three products are as below:

Sr. No.	*Details*	*Alpha*	*Beta*	*Gamma*
1.	Selling price (Rs./unit)	100	125	115
2.	Variable cost (Rs./unit)	45	59	55
3.	Raw material requirement (kg/unit)	5	3	4
4	Production per year	15,000	13,500	14,200
5	Total raw material requirement	75,000	40,500	56,800

The fixed cost has been assessed as Rs. 22,40,000.00. As an engineer advise the management the following.

1. What should be the product mix which will consume the total raw material available?
2. What will be the total contribution from the above product mix
3. What will be the total annual profit?

Solution:

Sr. No.	*Details*	*Alpha*	*Beta*	*Gamma*
1.	Selling price (Rs./unit)	100	125	115
2.	Variable cost (Rs./unit)	45	59	55
3.	Contribution (Rs./unit)	55	66	60
4.	Raw material requirement (kg/unit)	5	3	4
5.	Contribution per kg of raw material	11	22	15
6.	Priority	III	I	II

Ans. 1

The product mix will be Beta : Gamma : Alpha :: 13,500 : 14,200 : 10,540

Ans. 2 & 3

Sr. No.	*Details*	*Alpha*	*Beta*	*Gamma*	*Total yearly contribution*	*Ans*
1.	Contribution (Rs./unit)	55	66	60		
2.	Production per year	10,540	13,500	14,200		
3.	Total contribution, product-wise per year	5,79,700	8,91,000	8,52,000		
4.	**Total Yearly Contribution**				23,22,700	2
5.	Fixed cost				22,40,000	
6.	**Profit**				82,700	3

1.2.2.2 Selling Price of the Product

Marginal contribution can also be used to finalize the selling price of the product. The following illustration clarifies the point.

Illustration 2: A domestic company is manufacturing some domestic product like fans. The total yearly production is 10,000 units. The details of cost are as below:

Sr. No.	*Details*	*Cost per unit (INR)*
1.	Direct materials	1,500.00
2.	Direct labor	500.00
3.	Total variable cost	2,000.00

The company has incurred a fixed cost of INR 20,00,000.00 and want to earn profit during the current year of 15,00,000

Solution:

Contribution required = Fixed cost + Profit envisaged
= 20,00,000.00 + 15,00,000.00
= 35,00,000.00

Contribution per piece = 35,00,000/10,000 = 350.00

Sale price = Contribution per piece + Variable cost/unit
= 350.00 + 2000.00 = 2350.00

1.2.2.3 *Make or Buy Decision*

The concept of marginal costing is also helpful in taking decision for the management, particularly in the case of make or buy decision. In the present situation it is very much required because it is not always advisable to manufacture within the organization. It has been explained through the following illustration.

Illustration 3: Zenith Equipment Ltd manufactures four components, the cost particulars are given below:

Elements of cost	*Components*			
	Component A	*Component B*	*Component C*	*Component D*
Direct materials	80	100	100	120
Direct labor	20	25	25	30
Variable overhead	10	12	15	10
Fixed overhead	15	23	20	20
Total	125	160	160	180
Output per m/c Hr (unit)	4	2	3	3

You are required to advise the management as to whether they should continue to produce all or some of these components they should buy them from a supplier who has quoted the following price"

A – 115.00; B= 175.00; C = 135.00; D = 185.00

Solution:

Elements of cost	*Components*			
	Component A	*Component B*	*Component C*	*Component D*
Direct materials	80	100	100	120
Direct labor	20	25	25	30
Variable overhead	10	12	15	10
Marginal cost per unit	110	137	140	160
Purchase price per unit	115	175	135	185
Excess of purchase price over marginal cost	5	38	-	25
Excess of marginal cost over purchase price	-	-	5	-

Component C should be purchased from the outside supplier because purchase price is less than the marginal contribution. Company should continue to manufacture components in order of A, D and B.

1.2.2.4 Entry into Foreign Market

There are four important strategies generally followed in the export market. They are:

1. **Skimming Pricing Strategy**

   In this strategy, the company will start with very high price to take care of R & D cost and other promotional costs. Then after sometime they will reduce the price to maintain the market.

2. **Penetration Pricing Strategy**

   The company starts with very low price in order to drive away all other competitors. When the market settled, they can increase the price.

3. **Marginal Cost Pricing**

   The total variable cost (material, labor and variable overhead) is also known as marginal cost. In order to compete in the export market, the company may decide its selling price as marginal cost. They should not consider any contribution from those export products. The fixed cost should be recovered from domestic market.

1.2.2.5 Competitor's Pricing

Under this method, pricing strategy of price leader is taken into consideration while arriving at the pricing decisions. A price leader is the firm, which initiates the price trends in the market.

1.2.2.6 Short Term Policy during Trade Depression

During the time of trade depression, the demand of products comes down and there is less chance of getting the products sold. Marginal cost technique helps the management to come out of this problem. The management should find out the marginal cost (material, labor and variable overhead) of the product and if the price is above this the product can be sold. The amount above the marginal cost will take care of the reduction of pressure of fixed cost. The marginal cost technique can also help the management to take decision on the units of production to meet the situation. The following illustration clarifies the concept.

Illustration 4: Misa Ltd has two factories, X and Y. Both produce the same product whose selling price is INR 150 per unit. The other details are as follows:

Items	*Factory X*	*Factory Y*
Capacity in units	10,000	15,000
Variable cost per unit	100	130
Fixed expenses	3,00,000	2,10,000

Due to trade depression, the demand is likely to be only 12,000 units. How much units should be produced in each factory?

Solution: The variable expenses are directly connected with the production. If there is more production there will be more expenses. Hence, the production should be carried out in the factory where the variable cost per unit is less.

The variable cost of Factory X is less than that of Factory Y. Hence out of 12,000 units, 10,000 units should be produced from X and the remaining 2,000 units should be produced form Factory Y.

1.2.2.7 Continuation or Deletion of Products

The continuation or deletion is a part of the business. The management requires some technique which can help them to take decision for the benefit of the company. The management can take resort to marginal contribution technique. The following illustration will clarify the point.

Illustration 5: M/s Zenith Ltd is manufacturing three products, namely, White, Black and Green. The management received information from finance using absorption costing that "Green" is at loss. The details of the products are as below:

Items	*White*	*Black*	*Green*
Sales	30,00,000	15,00,000	9,00,000
Cost of goods sold			
Variables	18,00,000	10,00,000	6,50,000
Fixed	5,00,000	2,50,000	1,50,000
Total	23,00,000	12,50,000	8,00,000
Selling expenses			
Variables	2,00,000	1,20,000	80,000
Fixed	1,50,000	75,000	45,000
Total	3,50,000	1,95,000	1,25,000
Profit (+)/Loss (–)	3,50,000	55,000	(25,000)

As an engineer, advise the management what they should do: continue the product or delete the product.

Solution: Let us look to the problem using marginal contribution technique. The details have been put as below:

Items	*White*	*Black*	*Green*
Sales	30,00,000	15,00,000	9,00,000
Variables costs			
Production	18,00,000	10,00,000	6,50,000
Selling expenses	2,00,000	1,20,000	80,000
Marginal Cost	20,00,000	11,20,000	7,30,000
Contribution	10,00,000	3,80,000	1,70,000

It is cleared that "Green" has a positive contribution margin. Hence, it should not be deleted.

1.2.2.8 Acceptance/Rejection of Additional Volume at Reduced Price

Many times the situation arises that the management has to take decision either to take or not to take offer at the reduced price. This situation offers from the government. The management should be clear about the situation. There is a chance that accepting the order at reduced price, the company will be at loss. The marginal concept will come to rescue. The following illustration will elicit the point.

Illustration 6: M/S Engineering Company can manufacture 1000 units Product A annually. In the year 2017 they have produced only 400 units. The marginal cost of the unit is INR 300 and its selling price is INR 600. The fixed overhead is INR 1,50,000 for the year.

If a special order for 100 units of the product is obtained at INR 580 per unit, should the order be accepted?

Solution:

Items	*Detail*	*INR*
Fixed overhead		1,50,000
Less contribution of 400 units	400 units @300 = 1,20,000	
Less contribution of special order of 100 units	100 units @280 (580 – 300) = 28,000	
	Total 1,48,000	
Loss for the year		2000

At the selling price of INR 580, the order should not be accepted because the company is going to be at loss of INR 2000 annually.

1.2.2.9 Meeting Competition

In a business scenario, there will be always a situation when the same product is manufactured by more than one company. If you look into the situation there will be a total market may be domestic as well as international. This market requirement has to be taken by many companies. To increase the market share there will always be competition. Every company can innovate some new concept to attract the customer. Since cost is a big element of motivation, there will be a tendency to reduce the selling price.

There may be two situations; the first one is that because of reduction of sale value per unit, the revenue can increase through increase of sale and the increase of profit of the company. The second situation is that the company can increase production for keeping the profit same. The following illustration shows the second situation.

Illustration 7: The ABC Company produces and markets its products. Because of competition the company proposes to reduce the selling price. If the present profit is to be maintained indicate the number of units to be sold if the proposed reduction in selling price is 5%; 10% and 15%.

The following additional information is as below:

Particulars	*INR*	*INR*
Present sales turnover (30,000 units)		3,00,000
Variable cost (30,000 units)	1,80,000	
Fixed cost	70,000	2,50,000
Net profit		50,000

Solution:

Particulars	*Present price*	*Price reduction*		
		5%	10%	15%
Sales value per unit	10.00	9.50	9.00	8.50
Less variable cost	6.00	6.00	6.00	6.00
Contribution per unit	4.00	3.50	3.00	2.50

Total contribution = Fixed cost + Net profit = 70,000 + 50,000 = 1,20,000

Particulars	*Total contribution*	*Contribution per unit*	*Production volume (Total contribution/contribution per unit)*
Present price	1,20,000	4.00	30,000
Reduction by 5%	1,20,000	3.50	34,286
Reduction by 10%	1,20,000	3.00	40,000
Reduction by 15%	1,20,000	2.50	48,000

1.2.2.10 Suspension or Continuance of the Business

The business is not static. It varies based on different factors. The factors may be domestic as well as international. It is responsibility of the management to look into different aspects whenever there is a change or change is expected. As an example, one can think of the price of crude oil, which perhaps is the most vital element for industry. This item varies based on any situation of the world market, may be political, social or financial.

In the following illustration, it has been shown how the profit reduces over the years.

Illustration 8: Techno Engineering Company is manufacturing a product which they started long back. The details of the cost for the last three years are given below. The management feels that the company is not performing as it should, because there is an increase of fixed cost over the years. The details of the last three years are as below:

Items	*Year 2016*	*Year 2017*	*Year 2018*
Sales volume	30,000	20,000	20,000
Sales per unit	25	25	25
Variable cost	15	15	15
Fixed cost	1,50,000	1,65,000	1,95,000

The management is interested to know whether they should continue this business.

Solution:

Items	*Year 2016*	*Year 2017*	*Year 2018*
Sales volume	30,000	20,000	20,000
Sales per unit	25	25	25
Variable cost	15	15	15
Fixed cost	1,50,000	1,65,000	1,95,000
Total sales	7,50,000	5,00,000	5,00,000
Marginal cost	4,50,000	3,00,000	3,00,000
Marginal contribution	3,00,000	2,00,000	2,00,000
Profit/loss	**1,50,000**	**35,000**	**5,000**

The management should be advised to discontinue, till the market revives.

https://blog.blackcurve.com/how-to-price-a-product-in-a-competitive-market

1.2.3 Test On Chapter Two (Marginal Costing)

1.2.3.1 State Which of the Following Statements are True or False

1. Marginal costing technique is not meant for decision making.
2. Marginal cost is the summation of all variable costs.
3. Marginal contribution is not the difference of sales value and the marginal cost.
4. Profit/loss is calculated between the difference of marginal contribution and fixed cost.
5. Marginal cost includes prime cost and fixed cost.
6. An increase in the volume of production will reduce the variable cost per unit.
7. When fixed cost is deducted from the total cost one gets marginal cost.
8. Marginal costing can help management to take make or buy decision.
9. Product mix based on certain constraints can be taken through marginal contribution.
10. There is no difference between absorption costing and marginal costing.

1.2.3.2 What's Your Choice?

1. What is variable cost in a product?
   (a) The cost of the materials of the product
   (b) The labor and material cost of the product
   (c) Those costs which have direct relationship with the volume of production
2. In marginal costing technique
   (a) The fixed overhead has been divided into variable and fixed overhead
   (b) There is only fixed overhead and no variable overhead
   (c) Marginal cost is nothing but total cost
3. The most important element in marginal costing technique is,
   (a) Fixed cost
   (b) Marginal contribution
   (c) Total cost
4. Can marginal costing technique help management in make or buy decision?
   (a) Yes
   (b) No
   (c) Not sure
5. What is a marginal contribution?
   (a) Difference between sale value and total cost
   (b) Difference between sale value and prime cost
   (c) Difference between sales value and marginal cost
6. Contribution margin is also known as
   (a) Marginal income
   (b) Gross profit
   (c) Net income
7. Under the marginal costing system, the contribution margin discloses the excess of
   (a) Revenue over fixed cost
   (b) Variable cost over fixed cost
   (c) Revenue over variable cost
8. Is stock valued at variable cost in marginal costing system?
   (a) Yes
   (b) No
   (c) Not sure

9. Is marginal cost defined as
   (a) The summation of prime cost and fixed cost
   (b) The summation of prime cost and variable overhead
   (c) The summation of variable and fixed overhead
10. Marginal costing is also known as
   (a) Out-of-pocket costing
   (b) Historical costing
   (c) Indirect costing

Answers

True or false

1. (F)	2. (T)	3. (F)	4. (T)	5. (F)
6. (F)	7. (T)	8. (T)	9. (T)	10. (F)

What's your choice?

1. (c)	2. (a)	3. (b)	4. (a)	5. (c)
6. (a)	7. (c)	8. (a)	9. (b)	10. (a)

1.2.4 Problems

Problem 1

The annual sales volume of a product is 10,000 units. The details of the costs are as below:

Direct material cost = Rs. 12,000.00

Direct labor cost = Rs. 8000.00

Variable overhead = Rs. 6000.00

What is the marginal cost per unit of the product?

Solution: Rs. 2.60

Problem 2

The fixed expense of an industry is Rs. 3,00,000.00. The variable cost per unit is as below:

Direct material cost = 12.00

Direct labor cost = 8.00

Variable overhead = 5.00

Selling price per unit 35.00

What is the marginal contribution per unit?

Solution: Rs. 10.00

Problem 3

Fill up the blanks. Each case is different than others. All the cases are producing one product

Case no.	*Unit sold*	*Sales (Rs)*	*Selling price (Rs)*	*Variable expenses (Rs)*	*Fixed expenses (Rs)*	*Profit or loss*
1.	Blank	50,000	10	Blank	10,000	10,000 (Profit)
2.	8,000	Blank	8	40,000	Blank	9,000 (Profit)
3.	3,000	45,000	15	Blank	18,000	3,000 (Loss)
4.	9,000	81,000	9	45,000	20,000	Blank

Solution:

Case 1 : 5000 units and variable expenses : 30,000.00

Case 2 : Sales 64,000.00 and fixed expenses 15,000.00

Case 3 : Variable expenses 30,000.00

Case 4 : Profit 16,000.00

Problem 4

Sagar Engineering Limited manufactures one product. The annual plan of production is expected to be 10,000 units. The company has a fixed cost of Rs. 20,00,000.00. The details of other expenses are as below:

Direct material cost is Rs. 100.00

Direct labour cost is Rs 40.00

The company wants a profit of Rs. 10,00,000.00. What should be the price of the product?

Solution: Rs. 440.00

Problem 5

Kamal Company produces a single article. Following cost data is given about its product:

Selling price per unit Rs. 40

Marginal cost per unit Rs. 24

Fixed cost per annum Rs. 16,000

Calculate:

(a) Sales to earn a profit of Rs. 2,000, and (b) Profit at sales of Rs. 60,000

Solution: (a) 1125 units; (b) Rs. 8000.

Problem 6

Hitech Company manufactures a single product having a marginal cost of Rs. 1.50 per unit. Fixed cost is Rs. 30,000 per annum. The market is such that up to 40,000 units can be sold at a price of Rs. 3.00 per unit, but any additional sale must be made at Rs. 2.00 per unit. Company has a planned profit of Rs. 50,000. How many units must be made and sold?

Solution: 40,000 (Old) + 40,000 (New) = 80,000 units

Problem 7

Automatic Industry Limited, manufactures three products. The following is the cost data relating to products A, B, and C.

Figs. in Rs.

Products	*A*	*B*	*C*	*Total*
	Rs.	*Rs.*	*Rs.*	*Rs.*
Sales	1,50,000	90,000	60,000	3,00,000
Variable cost	1,20,000	63,000	36,000	2,19,000
Contribution	30,000	27,000	24,000	81, 000
Fixed cost		40,500		
Profit		40,500		

Prove that how knowledge of marginal costing can help management in changing the sales mix in order to increase profit of the company.

Solution: Keeping total production same, company should change the sales mix in a way that emphasis should be on producing product C and B.

Hint: Find out the P/V Ratio of A, B, C. C is the highest, then B and then A.

Problem 8

A company has a machine No. 9 which can produce either product A or B. The cost data relating to machine A and B are as follows:

Particulars	*Product A*	*Product B*
Selling price	Rs. 20.00	Rs. 30.00
Variable expenses	Rs. 14.00	Rs. 18.00
Contribution	Rs. 6.00	Rs. 12.00

Additional information:

(a) Capacity of machine No. 9 is 1000 hr.

(b) In one hour machine No. 9 can produce 3 units of A and 1 unit of B.

Which product should machine No. 9 produce?

Solution: Machine 9 should manufacture Product A.

Hint: Use contribution of products.

Problem 9

Smith & Company Ltd. has three divisions. Each division makes a different product. The budgeted data for the next year is as follows: Divisions

Division	A	B	C
	Rs	*Rs*	*Rs*
Sales	1,12,000	56,000	84,000
Direct material	14,000	7,000	14,000
Direct labor	5600	7000	22,400
Variable overhead	14,000	7000	28,000
Fixed cost	28,000	14,000	28,000
Total cost	61,600	35,000	92,400

The management is considering closing down division C. There is no possibility of reducing variable costs. Advise whether or not division C should be closed down.

Solution: Division C should be closed. It incurs loss.

Problem 10

The cost analysis of two products A and B is given below:

Particulars	*Product A*	*Product B*
	Rs.	*Rs.*
Material Rs. 2.50 per unit	25	45
Labor @ Rs. 1 per hour	12	---
Labor @ Rs. 1.50 per hour	---	15
Variable overheads	2	5
Selling price	70	80

On the basis of the above information, which product would you recommend to be manufactured if labor is the key factor and if material is the key factor?

Solution: If labor is the key factor then product B and if material is the key factor then product A should be produced

1.3 BREAK-EVEN POINT (BEP) ANALYSIS

1.3.1 Introduction

The management of any enterprise is always interested to know how the business flourishes. How you can improve profit on one hand at the same time how one can get out of the loss. In order to take necessary action, marginal costing technique can become a great help. This technique can indicate what the present situation is and what should be your decision in both the situations. In order to do that I bring back the Table 2 and illustrate a graphical model which can be used for any future situation. I repeat the assumptions which are necessary to achieve both the situations above.

Break-even point analysis is also known as cost-volume profit analysis.

Assumptions for marginal costing:

1. The fixed cost and the variable cost should not be changed during the period under review.
2. The product mix should remain fixed.
3. The sale value of the product should remain same during the period under review
4. There should not be any inventory during the period.

Concept of marginal cost has been put again in order to clarify certain points as below:

Sr. No.	*Elements of cost*	*Cost of 100 units*	*Cost of 101 units*	*Increase of cost*
1.	Direct material	50.00	50.50	0.50
2.	Direct labor	30.00	30.30	0.30
3.	Variable overhead	20.00	20.20	0.20
4.	Marginal cost (Sum of 1, 2, 3)	100.00	101.00	1.00
5.	Fixed cost	1,800.00	1,800.00	Nil

The concept of the above data can be put in the graphical model which can help to analyze the problem in any situation. This is shown in Exhibit 1

The X-axis represents the production and the Y-axis, the cost. Since the fixed cost remains same for all the productions, so a horizontal line is drawn parallel to X-axis. Marginal cost line or variable cost line starts from the origin because when there is no production there is no cost. To produce one unit, the variable cost is Re. 1.00, hence the slope of the curve will be 45° (tan θ = 1.00). A parallel line of variable cost line is drawn from the intersection of fixed line and the Y-axis. This new line will indicate the total cost line. If a vertical line is drawn on any production no, the from the graph one can get the value of Fixed cost; Variable cost and obviously the total cost against that production quantity.

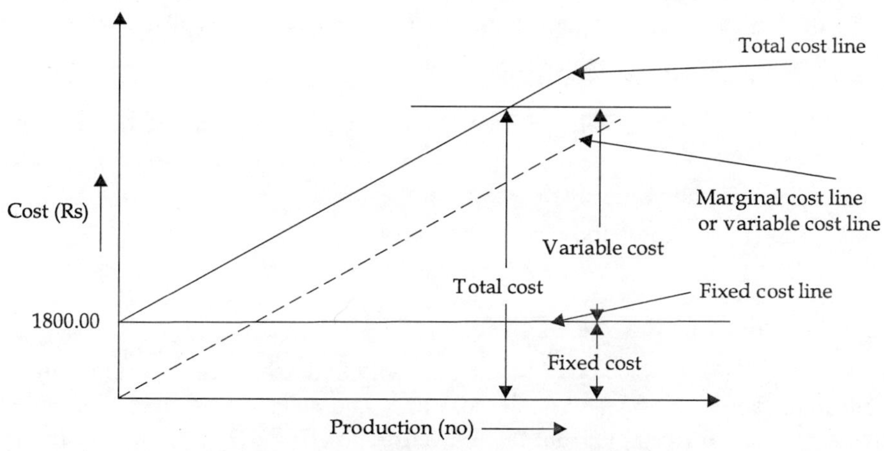

Exhibit 1 Sales Price Increased

Exhibit 2 has been drawn taking into account two major lines, namely, total cost line (TCL) and fixed cost line (FCL). If the price of the product is Rs. 19.00 per unit, then a line can be drawn in the same graph. The line will start from the origin; because when there is no sale, there will not be any sale value. It can be observed that sales line intersects the total cost line. This point of intersection is very important for the decision making for the management. This point is known as Break-Even Point (BEP)

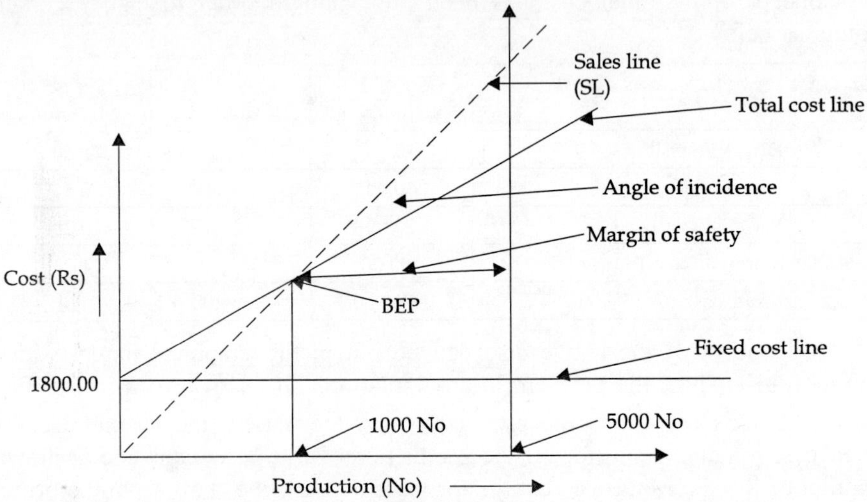

Exhibit 2

In the above example, BEP is coming as 1000 units. If the company is now manufacturing 5000 units, a vertical line can be drawn at 5000 units, the distance between BEP and the vertical line is called Margin of Safety. More the distance more is the stability of the company.

The angle at the BEP between sales line and total cost line is called the angle of incidence. If the angle is more, the profit element will be better.

1.3.2 Characteristics of BEP

1. This point where sale revenue is equal to total cost means no profit no loss.
2. The left hand side of this point is the loss zone.
3. The right side of the point is the profit zone.
4. The profit will start early if management can reduce the BEP.

1.3.3 Effect of Change of Lines

If the EXHIBIT 2 is analyzed, then one will find that there are three lines. Those are fixed cost line, variable cost line, and sales line. These three lines are responsible for bringing out break-even point (BEP).These three lines have effect on BEP. The effects of changing lines is discussed below.

1.3.3.1 Sales Line

If the sale price per unit changes, then there will be a change in the sales line itself. The EXHIBIT 2 will be redrawn below and there will be an imposition of the change of sales line. This will be mentioned as EXHIBIT 3 (when sales price per unit is increased) and EXHIBIT 4 (when sales price per unit decreases).

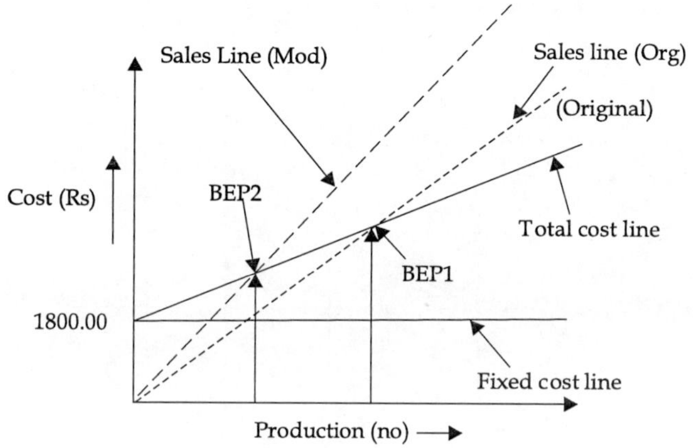

Exhibit 3: Sales Price Increased

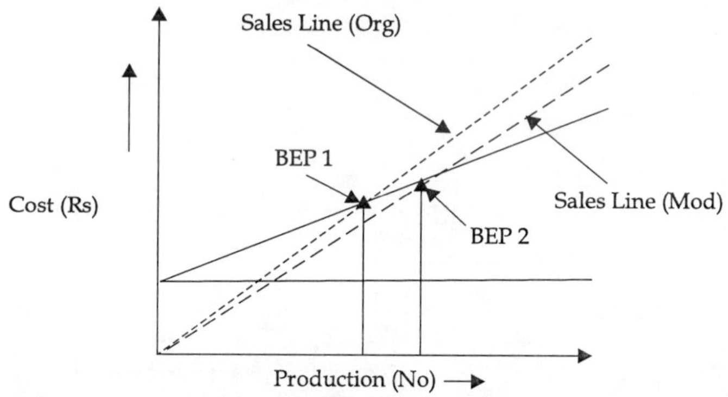

Exhibit 4: Sales Price Reduced

In EXIHIBIT 3, the sales price is increased from the present level. The effect of this is that BEP has been reduced; meaning thereby is that the management starts profit earlier than what was in the previous situation. But in the present scenario the price is fixed based on what the market can bear, hence there is less chance to increase sales price in normal situation.

In EXIHIBIT 4, the sale prices per unit has been reduced. Because of this there will be a change in the slope of the sales line. Now it will intersect the total cost line. Now the point of intersection is shifted on the right side. This means the BEP will increase. This means profit will start later than the previous situtation. This is the situation no management will like to have.

Management can intentionally reduce the sales price on some particular situation like trade depression or entrance into export market.

1.3.3.2 Variable Cost Line

The variable cost includes direct material cost, labor cost and variable expense. EXIHIBIT 5 (Increase of variable cost line) & EXHIBIT 6 (Decrease the variable cost line). are shown below:

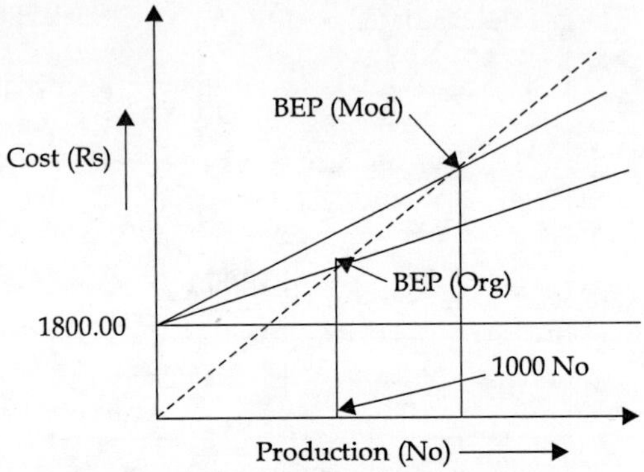

Exhibit 5: Increase variable cost line

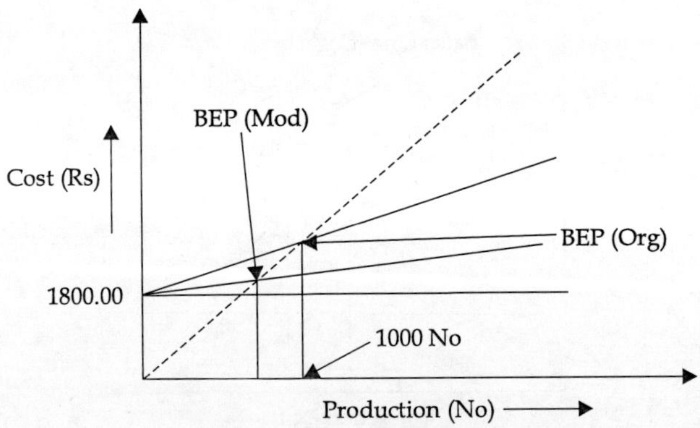

Exhibit 6: Decrease variable cost line

In EXHIBIT 5, because the variable cost has increased, so the variable cost line's slope has also increased. This new line has intersection with sales line which has shifted towards the right side. It indicates that BEP has been shifted towards the higher value. This means the profit will start late, not an acceptable situation.

In EXHIBIT 6, the variable cost has been reduced. The new line intersects the sales line earlier than the present case. Hence, the profit will start earlier. This situation is what every management wants. The reduction of variable cost can be achieved through various initiatives.

The variable cost includes, direct material cost, direct labor cost and variable expenses, which has been taken out from the fixed overhead.

The direct material cost can be reduced by doing the following actions.

1. Because of the technology change there is a lot of new materials which can be used to reduce the material cost.
2. Change of design is possible because of new materials; this may reduce weight of the product and thereby reduce the material per unit.
3. New design can help to eliminate some of the components which will help to reduce material cost.
4. Re-engineering the product will also help to reduce the direct material cost.
5. Use standard items instead of non-standard items.

The direct labor cost can also be reduced using different initiatives available.

1. New high efficiency machines like CNC machines can replace the old machines. This will reduce the time of manufacturing the product.
2. Skilled operators help to improve the quality by reducing the rejection/ rectification.
3. New processes of heat treatment improve the quality of product which will reduce the labor cost.

The variable expenses can also be reduced using the new concepts available.

1. Use solar energy to illuminate the working place. This is possible because a new type of overhead shade material is available.
2. The tool cost can be reduced by using the latest types of tools whose product life is high.
3. Compressed air can be properly reduced by using suitable gadgets.

1.3.3.3 *Fixed Cost Line (FCL)*

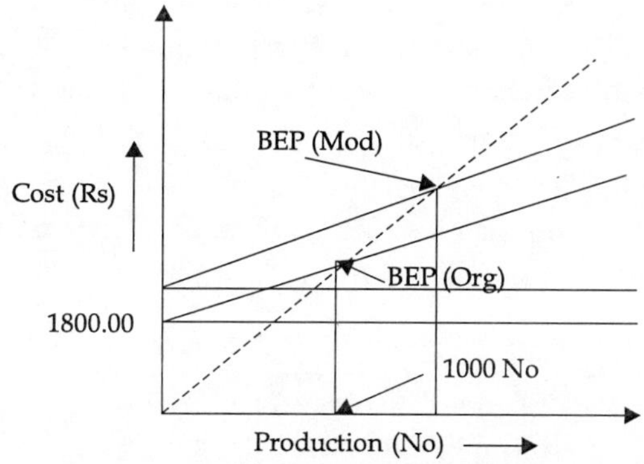

Exhibit 7: Increase FCL

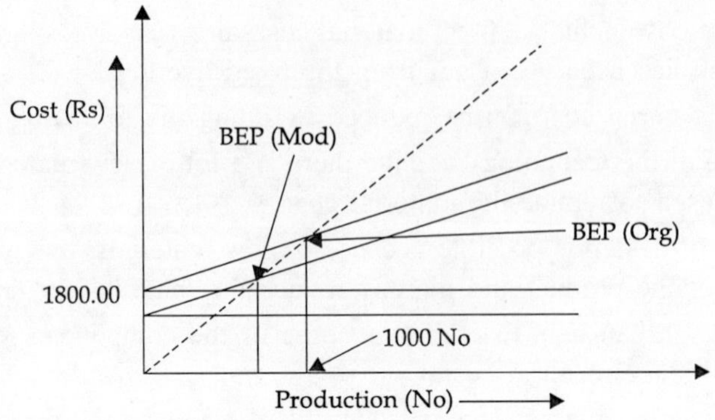

Exhibit 8: Decrease FCL

In EXHIBIT 7, fixed cost has been increased. Because of that the fixed cost line change its position upwards. Now the total cost line has shifted towards the right side. This indicates that BEP has more value, that means, profit starts late. No management is interested in this type of situation. Hence, the management always will try to keep the fixed cost unchanged if it could not be improved.

In EXHIBIT 8, the fixed cost has been reduced. It will have an effect of reducing the BEP. This means that profit will start earlier than the original case. Management is always interested to reduce the fixed cost.

The fixed cost is an area which should be improved to make the organization competitive. As example, we can discuss some of the initiatives and identify some area where improvement can be done.

1. The management can use Pareto diagram to find out the overheads of "A" category. Maximum overhead cost should be taken for improvement.
2. Electricity cost is always high, because of the outdated gadgets. The management should resort to the latest innovative items. Instead of using incandescent lamp or fluorescent tube, shift to LED lamp. There may be some initial cost but ultimately there will be gain for the management.
3. Compressed air is an important element in many industries. Proper gadgets help to reduce unnecessary wastage of compressed air.
4. Tool cost is also an item of high cost. Proper training of the operators can improve the cost of tool consumption.
5. Lastly, the operators at all levels should be aware of the cost of all items. This awareness will help to reduce the cost of fixed cost.

1.3.4 Profit Cost Volume Ratio

This is also known as P/V ratio. This concept is useful to find out the break-even point (BEP). Here profit means contribution and volume is the production volume. The mathematical equation will be as below:

P/V Ratio = Contribution/Sales
= (Fixed cost + Profit)/Sales
= (Fixed cost – Loss)/Sales

1.3.5 Some Worked out Examples

Problem 1

Hitech Company is manufacturing three products X, Y and Z. The details are given below. All Figs are in INR

Details	*Product X*	*Product Y*	*Product Z*
Selling price per unit	50	40	30
Variable cost per unit	20	15	10
Fixed cost	210.000	120.000	80,000

Calculate break-even quantity of X, Y and Z.

Solution:

Details	*Product X*	*Product Y*	*Product Z*
Total fixed cost	210,000	120,000	80,000
Selling price per unit	50	40	30
Variable cost per unit	20	15	10
Marginal contribution per unit = (Sale price – Variable cost)	30	25	20
Break even quantity = (Fixed cost/marginal contribution per unit)	7000	4800	4000

Problem 2

Find the selling price per unit on the basis of the following information.

Variable cost per unit = Rs. 7.20

P/V ratio = 40%

Solution:

P/V Ratio = [Contribution/Sales] × 100

Contribution = Sales – Variable cost

P/V ratio = [(Sales – Variable cost)/Sales] × 100

P/V ratio/100 = 1 – Variable cost per unit/Sales per unit

Variable cost per unit/Sales per unit = 1 – 40/100 = 60/100

Sales per unit = (Variable Cost per unit/60) × 100 = (7.2 × 100)/60 = 12.00

Problem 3

M/s X Ltd is manufacturing a single product. The sale of the product is 1,50,000 units. The details of the cost structure are as below:

Particular	*Amount (Rs.)*	*Amount (Rs.)*
Sales of 1,50,000 @ 10.00		**15,00,000**
Cost of Sales		
Direct material cost per unit	4.00	6,00,000
Direct labour cost per unit	1.00	1,50,000
Production overhead (variable) per unit	0.50	75,000
Production overhead (fixed) per unit	2.00	3,00,000
Administrative overhead (fixed) per unit	1.00	1,50,000
Selling and distribution overhead (variable) per unit	0.20	30,000,
Selling and distribution overhead (fixed) per unit	0.40	60,000
Total		**13,65,000**
Profit		**1,35,000**

In order to improve the performance management has engaged a consultant. The consultant has reduced the direct material cost per unit to 3.50 by changing the new material. They have also reduced production overhead (fixed) by introducing new lighting system to 1.50 per unit.

Advise the management whether the new proposal is accepted based on BEP Analysis.

Solution:

Particulars	*Existing (Rs.)*	*Proposed (Rs.)*
Variable Cost (per unit)		
Direct material cost	4.00	3.50
Direct labor cost	1.00	1.00
Production overhead	0.50	0.50
Selling and distribution overhead	0.20	0.20
Sub-total	**5.70**	**5.20**
Fixed Cost (per unit)		
Production overhead	2.00	1.50
Administrative overhead	1.00	1.00
Selling and distribution overhead	0.40	0.40
Sub-total	**3.40**	**2.90**

Let N_e be the BEP of the existing and N_p is the BEP of the proposed.

Existing:

$$N_e \times 10.00 = N_e \times 5.70 + 3.40 \times 1,50,000$$

or $$N_e (10.00 - 5.70) = 5,10,000$$

or $$N_e = 5,10,000/4.30$$

$$= \mathbf{1,18,605\ (BEP)}$$

Proposed:

$$N_p \times 10.00 = N_p \times 5.20 + 2.90 \times 1{,}50{,}000$$

or $$N_p\,(10.00 - 5.20) = 4{,}35{,}000$$

or $$N_p = 4{,}35{,}000/4.80$$

$$= \mathbf{90{,}625\ (BEP)}$$

Since the BEP has been reduced, the management will start profit early. Hence, the proposal should be accepted.

Problem 4

M/S XYZ produces three products, Alpha, Beta and Gamma. All the three products require the same raw material. The yearly material availability is only 1,50,000 kg. The details about the three products are as below:

Sr. No.	*Details*	*Alpha*	*Beta*	*Gamma*
1.	Selling price (Rs./unit)	100	125	115
2.	Variable cost (Rs./unit)	45	59	55
3.	Raw material requirement (kg/unit)	5	3	4
4.	Production per year	15,000	13,500	14,200
5.	Total raw material requirement	75,000	40,500	56,800

The fixed cost has been assessed as Rs. 22,40,000.00. As an engineer please advise the management the following.

1. What should be product mix which will consume the total raw material available?
2. What will be the total contribution from the above product mix
3. What will be the total annual profit?

Solution:

Sr. No.	*Details*	*Alpha*	*Beta*	*Gamma*
1.	Selling price (Rs./unit)	100	125	115
2.	Variable cost (Rs./unit)	45	59	55
3.	Contribution (Rs./unit)	55	66	60
4.	Raw material requirement (kg/unit)	5	3	4
5.	Contribution per kg of raw material	11	22	15
6.	Priority	III	I	II

Ans. 1

The product Mix will be Beta: Gamma: Alpha :: 13,500:14,200:10,540

Ans. 2 & 3

Sr. No.	*Details*	*Alpha*	*Beta*	*Gamma*	*Total yearly contribution*	*Ans*
1.	Contribution (Rs/unit)	55	66	60		
2.	Production per year	10,540	13,500	14,200		
3.	Total contribution, product-wise per year	5,79,700	8,91,000	8,52,000		
4.	**Total Yearly Contribution**				23,22,700	**2**
5.	Fixed cost				22,40,000	
6.	**Profit**				82,700	**3**

Problem 5

From the following information find out:

(a) P/V ratio

(b) Sales &

(c) Margin of safety

Fixed cost = Rs. 40,000

Profit = Rs. 20,000

BEP = Rs. 80,000

Solution:

(a) P/V Ratio

P/V ratio = (Fixed cost + Profit)/Sales

At BEP, profit is zero, sales at BEP is 80,000 and fixed cost is 40,000

So, P/V Ratio = 40,000/80,000 = 0.50 or 50%

(b) Sales

P/V ratio = (Fixed cost + Profit)/Sales

Or sales = (Fixed cost + Profit) P/V ratio

Or sales = (40,000 + 20,000) × 50%

Or sales = 60,000/0.50 = 1,20,000

(c) Margin of Safety

Margin of safety = Sales – Sales at BEP

= 1,20,000 – 80,000

= 40,000

1.3.6 Test on Chapter Two (Break-even Point)

1.3.6.1 State Which of the Following Statements are True or False

1. Break-even point is not an important element for management's decision.
2. P/V ratio can help to find out the BEP.
3. BEP changes if there is a change in fixed cost.

4. If the variable cost increases, the BEP will decrease.
5. One of the best ways to improve BEP is to reduce fixed cost.
6. Margin of safety indicates the position of the business.
7. The profit will not start early if the management reduces the BEP.
8. BEP is the intersection of sales line and total cost line.
9. All variable costs are not included in the marginal cost.
10. At breakeven point, margin of safety is nil.

1.3.6.2 What's Your Choice?

1. What are the elements required to calculate BEP?
   (a) Fixed cost, profit/loss and sales line
   (b) Profit/loss, variable cost and sales line
   (c) Fixed cost, variable cost and sales line
2. What are the items included in the variable cost?
   (a) Direct material cost, direct labor cost and variable overhead cost
   (b) Profit, direct material cost and direct labor cost
   (c) Variable overhead, profit and direct material cost
3. Will sales line start from
   (a) Intersection of fixed cost line with Y-axis
   (b) Can start from any point
   (c) Will start from the origin
4. P/V Ratio is equal to
   (a) Contribution/sales
   (b) Fixed cost/sales
   (c) Variable cost/sales
5. What is margin of safety?
   (a) Difference between present sales and the sales at BEP
   (b) Difference between sale value and prime cost
   (c) Difference between sales value and marginal cost
6. Will there be change if fixed cost is reduced or increased
   (a) No
   (b) Yes
   (c) Not sure
7. What is the intention of the management regarding BEP?
   (a) To start the profit earlier than the present situation
   (b) Management is not interested about BEP
   (c) It is just information, not necessary to act upon
8. The P/V ratio of a product is 0.4 and the selling price is Rs. 40 per unit. What will be the marginal cost of the product?

(a) Rs. 10
(b) Rs. 20
(c) Rs. 24

9. A large margin of safety indicates
   (a) Overproduction
   (b) The soundness of business
   (c) None of these
10. A decrease in the sales price will
   (a) Increase the BEP
   (b) Lower the net profit
   (c) Lower the BEP

Answers

True or false

1. (F)	2. (T)	3. (T)	4. (F)	5. (T)
6. (T)	7. (F)	8. (T)	9. (F)	10. (T)

What's your choice?

1. (c)	2. (a)	3. (c)	4. (a)	5. (a)
6. (b)	7. (a)	8. (c)	9. (b)	10. (a)

1.3.7 Problems

Problem 1

The amount of fixed expenses of a company is Rs. 3,60,000. The variable cost per unit is Rs. 23 and the selling price is Rs. 35. What is the BEP?

Solution: 30,000

Problem 2

A management consultant has given two proposals for the construction of a building. The details are as below.

Fig. in Rs. per square ft.

Item	*Proposal 1*	*Proposal 2*
Design	10,000	25,000
Bricks	12	12
Cement	16	15
Site	40,000	1,00,000
Steel	22	18

Calculate the quantity of space necessary to be constructed for the proposal to become cost effective using break-even analysis.

Solution: Beyond 17,000 square feet Proposal 2 will be profitable.

Problem 3

M/s X Ltd, manufactures shampoo which it fills in a standard size pack for marketing. A VE team has proposed some improvement because of which there is a reduction in variable cost at the cost of increase of fixed cost. The details are as below:

Item	*Existing*	*Proposal*
Selling price Rs.	25	25
Variable cost Rs.	15	10
	Fixed expenses	
Rent	8000	8975
Salaries	7000	7000
Others	5000	6000

Advise the management using break-even analysis.

Solution: BEP (existing) = 2000 and BEP (proposal) = 1465. Proposal is acceptable.

Problem 4

Calculate the sale value at the break-even point from the following details.

Total sales : INR 10,00,000

Variable expenses : INR 6,00,00

Net profit : INR 1,00,000

Solution: INR 7,50,000

Problem 5

Surya Industry earned a profit of Rs. 4,00,000. Sales of the industry is Rs. 25,00,000 and fixed expenses is Rs. 6,00,000.

Moonlight Industry earned a similar profit from a similar sale. But the fixed expense is Rs. 8,50,000.

Calculate the break-even point for both the industries.

Solution: Surya 15,00,000 and Moonlight 17,00,000

Problem 6

The following budget has been prepared at 70% level of home market:

Units = 4200

Wages = 12600

Materials = 21,000

Fixed cost = 7000

Variables cost = 2100

Total = 42,700

The selling price in India is Rs. 15. In Sri Lanka about 800 units may be sold only at Rs. 10 and in addition 25 paise per unit will be expenses of freight, etc. Do you advise trying for the market in the Sri Lanka?

Solution: It is advisable to try the Sri Lankan market at Rs. 10 per unit as by doing so there is an increase of Rs. 1000.

Problem 7

From the following data calculate:

(a) Break-even point expressed in the amount of sales in rupees.

(b) Number of units that must be sold to earn a profit of Rs. 60,000 per year

Items	*Rs*
Sale price	20 per unit
Variable manufacturing cost	11 per unit
Variable selling and distribution cost	3 per unit
Fixed factory overhead	5,40,000 per year
Fixed selling cost	2,52,000 per year

Solution: (a) Rs. 26,40,000; 1,42,000 units

Problem 8

A manufacturer produces 1500 units of various products annually. The marginal cost of each product is Rs. 960 and the product is sold for Rs. 1200. Fixed cost incurred by the company is Rs. 48,000 annually. Calculate P/V ratio and what would be the break-even point in terms of output and in terms of sales value?

Solution: P/V ratio is 20%; break-even point in terms of output is 200.
Break-even point in terms of value is Rs. 2,40,000

Problem 9

A company is providing its product to the consumer through the wholesalers. The managing director of the company thinks that if the company starts selling through retailers or to the consumers directly, it can increase its sales, charge higher prices and make more profit.

On the basis of the following information, advise the managing director whether company should change its channel of distribution or not:

Particulars	*Wholesaler*	*Retailer*	*Consumer*
	Rs.	*Rs.*	*Rs.*
Sales per unit	3.60	5.25	6.00
Estimated sales per year (units)	1,00,000	1,20,000	1,80,000
Selling and distribution overheads (per unit)	0.40	1.00	1.50

Cost of production: variable cost Rs. 2.50 per unit, fixed cost Rs. 50,000

Solution: Sales should be made directly to the consumers as this channel contributes higher profit.

Problem 10

From the following information. Calculate sales to earn profit of Rs. 2,40,000.

Particulars:

Sales = Rs. 8,00,000

Fixed cost = Rs. 3,60,000

Variable cost = Rs. 5,60,000

Solution: 20,00,000

2 Methods

2.1 COST MODEL

2.1.1 What is Cost?

In order to produce an item one has to employ resources. The amount of resource required to create an item is different from item to item depending upon various factors. But the types of resources are the same. Each resource is converted into monetary units, when one adds all the resources in monetary terms the sum of that is called the cost of that product.

Cost of the product is important because on the basis of this seller deals his product with buyer. The seller adds an item called profit to finalize his product which is known as PRICE. Hence, one can say that the nucleus of trading an item is COST.

2.1.2 Why "Cost" is Important?

To find the value of the product, the cost of the product is an important item. The second item for the value is known as function which is the purpose of the product. This function is converted into monetary value.

The mathematical formula for value is shown as below:

$$\text{Value} = \frac{\text{Function (in terms of monetary value)}}{\text{Cost (in monetary value)}}$$

Hence, it is clear that in order to make your product cost effective, one should know details about cost. The cost is made up of direct material, direct labor and overheads. It needs explanation for clarity.

Direct materials are those materials which make the product and have relation with the number of products. That is, if quantity increases/ decreases the requirement of the material increases/decreases.

Direct labor has the same relation as that of direct materials.

Overhead is the summation of indirect materials, indirect labor and indirect expenses. Some indirect expenses examples are shown below:

1. Canteen expenses
2. Consumable stores
3. Depreciation of plant, machinery and equipment
4. Depreciation of buildings

5. Heating, lighting, etc.
6. Fuel
7. General purpose tools
8. Insurance charges
9. Power
10. Rent, rates, etc.
11. Repair and maintenance of plant, machinery, factory, building, etc.
12. Sundry expenses
13. Supervisors' salary
14. Telephone
15. Taxes (municipal tax or tax imposed by local authority)

2.1.3 What is the Cost Model?

(a) A diagrammatic representation of cost based on a hierarchical structure [often work breakdown structure] of product, components or functions.
(b) A model which sums up to the total cost of the product or project.
(c) A financial representation of the total cost of families of systems, components or parts within a total complex product, system, structure or facility.
(d) The results of the models are typically necessary to obtain approval for budgets, and other financial planning.

2.1.4 Types of Cost Models

2.1.4.1 Cost Model of Product Overhead

Production overhead		*Administrative overhead*		*Selling overhead*		*Distribution overhead*	
Items of expenses	*Amount ($)*	*Items of expenses*	*Amount ($)*	*Items of expenses*	*Amount ($)*	*Items of expenses*	*Amount ($)*
Rent & taxes	12,000	Salary to staff	2000	Advertisement	5650	Packing materials	2500
Electricity	4600	Bank charges	650	Travelling allowance	2500	Warehouse rent	6200
Power	4000	Books & periodicals	5000	Commission to salesman	3700	Cost of manual handling	500
Machinery repairs	20,000	Furnishing	10,500	Market research	4300	Loading & unloading cost	3500
Depreciation	5000	Stationery	2000	Stationery	700	Delivery cost	2350
Fire insurance	1000	Legal fees	6000	Salary to executives	10,500	Insurance of finished goods	5200
Total	46,600		26,150		27,350		20,250

One can identify from the cost model which is the area where there is potential for improvement of cost.

2.1.4.2 Cost Model for Function Cost

Component	*Cost*	*Function (Active verb and Measurable noun)*								*Figs in $*	
		Contain liquid	*Produce shape*	*Reduce thermal conduction*	*Withstand load*	*Attract customer*	*Convey message*	*Accommodate finger*	*Reduce heat flow*	*Join components*	*Withstand thermal stress*
Body	215.00	12.00	10.00	10.00	20.00	95.00	43.00		10.00	5.00	10.00
Handle	35.00		5.00	5.00	10.00	2.00		1.00	5.00	2.00	5.00
Total	250.00	12.00	15.00	15.00	30.00	97.00	43.00	1.00	15.00	7.00	15.00
Benchmarking cost											
Target cost											

Adding benchmarking for similar products will help one to get the maximum from this type of cost model. Target cost can also be incorporated for increasing the market share through making the product cost effective.

2.1.4.3 Cost Model for Construction

Total Construction					
Site work ($)	*Building*				
	Structural ($)	*Architectural ($)*	*Mechanical ($)*	*Electrical ($)*	*Equipment ($)*
Site preparation	Foundation	Exterior closure	Plumbing	Services & distribution	Fixed assets
Site improvement	Special foundation	Roofing	HVAC	Lighting & power	
Site utilities	Sub structure	Interior construction	Fire protection	Special electricals	
Garage	Super structure	Conveying system	Special mechanical BMS	Misc. electricals	
Overhead & profit					
Construction cost at Launch =		Total construction + Contingency + Escalation			

Construction cost at launch = Total construction + Contingency + Escalation

Contingency is generally put as the percentage of total construction cost or based on the past performance of the similar construction.

Escalation figure will depend upon the environment prevailed during that period.

2.1.4.4 Cost Model for Infrastructure

Cost measured directly	*Cost ($)*	*Allowance*	*Cost ($)*	*Contingency*	*Cost ($)*	*Reserve*	*Cost ($)*
Pier foundation	94.845	It is generally on lump sum or in % of direct measured cost (20% or as may be decided)		It is a buffer to mitigate situation like earth-quake etc. (external)		This is held by the owner if there is change in scope in the project in future	
Pier wall foundation	410.667						
Pier cap beam	224.000						
Deck	723.33						
Total	1452.842		290.5684				

Minimum cost of the project = Cost measured directly + Allowance
Expected cost of the project = Minimum cost + Contingency
Maximum cost of the project = Expected cost + Reserve.

2.1.4.5 *Function–Cost–Worth Model (Pencil)*

Function		*Type*	*Allocated cost to each function*			*Method of achieving worth*	*Value index*	*Value gap*
Verb	*Noun*	*B/S*	*Basis*	*Cost (Rs)*	*Worth (Rs)*			
Make	Mark	B	Cost of lead	0.77	0.50	The cost of lime	1.54	0.27
Protect	Lead	S	Cost of wood	1.26	0.25	The cost of binding of lime	5.04	1.01
Indicate	Specification	S	Cost of printing	0.03	0.03	Printing on the box	1.00	0.00
Indicate	Manufacturer	S	Cost of printing	0.06	0.06	Printing on the box	1.00	0.00
Enable	Grip	S	Cost of manufacturing hex shape	0.30	0.06	Cost of making conical shape	5.00	0.24
Improve	Appearance	S	Printing on pencil	0.58	0.10	Appearance of the box	5.80	0.48
Total	Product cost			3.00	1.00		3.00	2.00

Cost of the product (pencil)

Cost of 5 pencils + rubber + sharpener + packet = 25
Rubber = 2
Sharpener = 3
Packet = 5

Subtotal = 10

Cost of 5 pencils = 25 – 10 = 15 or cost of one pencil= 3.00

Calculation of worth

Item taken as chalk box of 20 pieces

Cost = Rs 25.00

Box = 5.00

Cost of chalk (20) = 20; Cost/piece = 1.00

2.1.5 Test On Part B (Cost Model)

2.1.5.1 State Which of the Following Statements are True or False

1. Cost model is a diagrammatic representation of cost based on a hierarchical structure.
2. Cost of the creation is not important because it is not involved anywhere in the business.
3. Administrative overhead is not a part of cost model of product overhead.
4. Cost model in construction is an essential item for real estate business.
5. Cost model of infrastructure will be different from that of construction industry.

2.1.5.2 What's Your Choice?

1. What is cost?
   (a) It is same as price
   (b) Cost is nothing but a notional value
   (c) The amount of resource required to create an item
2. One of the items of cost model of infrastructure
   (a) Contingency
   (b) Product overhead
   (c) Architectural
3. Cost model is an essential item of cost analysis
   (a) No
   (b) Yes
   (c) Not sure

Answers

True or false

1. (T) 2. (F) 3. (F) 4. (T) 5. (T)

What's your choice?

1. (c) 2. (a) 3. (b)

2.2 COST ESTIMATES

2.2.1 Introduction

The final outcome of the process of cost estimation is the cost estimate. It is the most important feature for running the business. The correctness of the statements will help the industry to reduce their chance of overrun. The cost estimates is also used for budget.

The cost estimates are used in, production, constructions, projects and programs, etc. Those who are responsible for these jobs are called estimators. The titles of the cost estimators are building estimators, electrical estimators, quantity surveyor or cost engineer. The job of estimators is very skilled and in many cases they should be approved by the respective authorities to make their estimates authentic and approved by the Government and other agencies.

2.2.2 Places of Use

(a) In **cost engineering**, cost estimation is a basic activity. The fixed capital investment provides the physical facilities. The working capital investment is a revolving fund to keep the facilities operating.

(b) In **project management, project-cost-management** is a major functional division.

Cost estimating is one of three activities performed in project cost management.

(c) In **system, product, or facility acquisition planning**, a cost estimate is used to evaluate the required funding and to compare with bids or tenders.

(d) In **construction contracting**, a cost estimate is usually prepared to submit a bid or tender to compete for a contract award.

(e) In **facility maintenance and operation**, cost estimates are used to establish funding or budgets.

(f) In an attempt to manage liability risk, some firms avoid the use of the word estimate and instead refer to the estimate as an **"opinion of probable cost."**

2.2.3 Role of Different Agencies to Develop the Cost Estimates

2.2.3.1 Cost Engineering

The cost engineering in an industry is responsible to develop the cost estimates of the product. The major elements of the cost of a product are: direct materials cost, labor cost and the overhead cost. While direct materials and labor can be ascertained without any difficulty, but the absorption of overhead cost is an important aspect for the product. Cost engineers use different methods depending upon the nature of the industry. The following methods are used by them as per the situation.

2.2.3.1.1 Methods of Absorption of Overhead (First Method)

(a) Percentage of direct material cost
(b) Percentage of direct labor cost
(c) Percentage of prime cost
(d) Labor hour rate
(e) Machine hour rate
(f) Rate per unit of output

The differences are shown through an example in Table I.

Overhead (A) US $	*Direct material cost (B) US $*	*Direct labor cost (C) US $*	*Prime cost D = (A+B) US $*	*Labor hours (E) US $*	*Machine hour (F) US $*	*No of units (Nos)*
50,000	1,00,000	20,000	1,20,000	12,500	5,000	25,000
For absorption	50% on direct materials	250% of direct wage cost	41.67 % on prime cost	$4.0 per labor hour	$5.0 per machine hour	$2.0 per unit of output

Examples of Cost Estimates Based on the First Method

The company has received an order of 1200 units which they have to complete in one month. Detailed data is as below. Based on the Table I, calculate cost estimates of 1200 units of production.

Item	*US$/unit*	*Cost for 1200 units (US$)*
Direct material/unit	100	1,20,000
Labor hr/unit	5.0	6000
Labor cost/unit	6.0	7200
Machine hr/unit	3.0	3600

Cost Estimates Based on Different Methods

Different methods	*Direct material cost*	*Direct wages*	*Overhead factor*	*Overhead absorption*	*Estimated cost for 1200 unit*
Percentage on direct material cost	1,20,000	7200	50%	60,000	1,87,200
Percentage of direct labor cost	1,20,000	7200	250%	18,000	1,45,200
Percentage of prime cost	1,20,000	7200	41.67&	53,004	1,80,204
Labor hour rate	1,20,000	7200	US4.00	24,000	1,51,200
Machine hour rate	1,20,000	7200	US$5.00	18,000	1,45,200
Rate per unit of output	1,20,000	7200	US$2.00	2,400	1,29,600

2.2.3.1.2 Methods of Absorption of Overhead (Second Method)

In this method some factors are attributed to different headings, which are given below.

Basic overhead factor: 100% (or a value of 1) is recommended for basic overhead (worker benefits, buildings, utilities, maintenance).

- **Equipment factor:** 50% (or a value of 0.5) is recommended for assembly and manufacturing operations that require equipment or tooling (to cover amortized costs, consumables, and service)
- **Special operation/tolerance factor:** 25% (or a value of 0.25) is recommended for special case operations (laser welding, CNC machining) and/or tight tolerance requirements (due to increased cost of equipment and personnel, levels of scrap and inspection time)

The cost estimates based on the above considerations are shown through an example.

Example: Based on the following data, calculate the cost estimate of the frame assembly sub-assembly in different methods. The processes go through three operations and finally assembly and inspection.

Cost Details of Frame Assembly for (one day assembly line production)

	Operation 1	*Operation 2*	*Operation 3*	*Operation 4*	*Purchase materials & components*	*Total*
	Cut bar stock to length and prep for welding	Place cut pieces in jigs for the welder and remove	Tack and finish welding of prepared pieces in the jigs for sub-assembly	Inspection, cleanup, etc. for the 8 hour shift		
(a) Time to complete in hr/ frame	3.0	1.0	1.0	0.5		
(b) Labor rate ($/hr)	17.72	17.72	20.18	6.69		
(c)Labor cost = a × b						
(d) Basic overhead factor	1	1	1	1		
(e) Equipment factor	0.5	0.5	0.5	0.5		
(f) Special operation/ tolerance factor	0.25	0	0	0		
(g) labor/ equipment cost = c × (1+d+e+f)	$146.19	$44.30	$50.45	$8.38	$100	$349.32

Frame production cost estimate

= Σ (labor/overhead/equipment costs) + purchased materials/components cost

= $[53.16*(1 + 1 + 0.5 + 0.25) + 17.72*(1 + 1 + 0.5) + 20.18* (1 + 1 + 0.5) + 3.35*(1 + 1 + 0.5)] + $100

= $257.67 + $100

= $349.32

Source: Hourly rate based on information about Columbus OH averages for different job classifications available from the US Department of Labor (http://stats.bls.gov/ncs/home.htm, May 2010).

2.2.4 Test on Part B (Cost Estimate)

2.2.4.1 State Which of the Following Statements are True or False

1. Cost estimate is the most important feature for running a business.
2. Those who are responsible for cost estimation are called cost calculators.
3. One of the places of use of cost estimation is facility maintenance and operation.
4. One of the methods of absorption of overhead cost is percentage of indirect material cost.
5. Many firms instead of using the word estimate use "opinion of probable cost".

2.2.4.2 What's Your Choice?

1. What is cost estimate?
   (a) The final outcome of the process of cost estimation is cost estimate
   (b) Cost estimate is nothing but total cost
   (c) Cost estimate is only for the direct materials
2. Cost estimate is used in the following place
   (a) Maintenance of machine
   (b) Supply change management
   (c) Construction contracting
3. Many cases estimation should be approved by the government and other agencies.
   (a) No
   (b) Yes
   (c) Not sure

Answers

True or false

1. (T) 2. (F) 3. (T) 4. (F) 5. (T)

What's your choice?

1. (a) 2. (c) 3. (b)

2.3 PARETO DIAGRAM

2.3.1 Who was Pareto?

Vilfred Pareto (1848-1922) was an Italian economist and sociologist. He worked on the wealth distribution among the people.

2.3.2 What is his Principle?

Vilfred Pareto observed in his study in Europe during 1900 that eighty per cent of wealth is concentrated to the twenty percentage of people. The Pareto principle is also known as 80-20 principle. This is also called the "vital few and the trivial many" rule. Though it was basically meant for wealth distribution, but it has been observed that this principle is valid in various other areas also. Because of its usefulness it is widely used in business.

2.3.3 What are the Successful Areas?

The areas where the Pareto principles are being used successfully are:

1. **Inventory control:** It has been observed that 20 to 25 per cent items in the inventory cost is responsible for the 80% of the total inventory cost. Pareto analysis will identify and will help to improve that situation.
2. **Maintenance:** Only 20% of the machines' maintenance cost is responsible for the 80% cost of the total maintenance cost.
3. **Customer response:** If you take notice of the customer response, one will find that it will be for a few causes only.
4. **Training:** Pareto analysis can help you to identify the persons who need training as per the requirement of the management.
5. **Medical services:** Even one analyzes the national health service, it will be observed that 80% expenses are meant for 20% of the population.
6. **Accounts receivable:** 80% account receivable is from 20% of the customers.
7. **Manufacturing:** If one analyzes the manufacturing cost of any product, it will be observed that 20% of the components is responsible for the 80% of the cost.

2.3.4 Steps to make Pareto Diagram

Step No. 1

Decide in which area it can be used. Let us take "foundry defects" as an example.

Step No. 2

Finalize the categories of defects which are seen in metal casting process. It is considered a good practice to have less than 10 categories. In this case six categories have been chosen which are important in the study as shown below:

- Cracks
- Surface scar
- Incomplete
- Blow hole
- Wrapage
- Others

Step No. 3

Against each category collect the frequencies against each category. It can be seen as below: Table 2.1

Table 2.1

Sr. No.	*Category of defects*	*Frequency*
1.	Surface scar	82
2.	Cracks	95
3.	Others	8
4.	Blowhole	20
5.	Warpage	15
6.	Incomplete	30

Step No. 4

A table is made putting defects in descending order. The provide code to each category. This is shown in Table 2.2.

Table 2.2

Sr. No.	*Code*	*Category of defects*	*Frequency*	*Cumulative frequency*	*Cumulative in %*
1	A	Crack	95	95	38.0
2	B	Surface scar	82	177	70.8
3	C	Incomplete	30	207	82.8
4	D	Blowhole	20	227	90.8
5	E	Warpage	15	242	96.8
6	F	Others	8	250	100
Total			**250**		

Step No. 5

Now draw a graph. The X-axis will be the defect code. The left Y-axis will be the frequency of the code. This will be a bar graph.

The right Y-axis considers the percentage of the code and a line graphs is drawn. This is shown in Fig. 2.1.

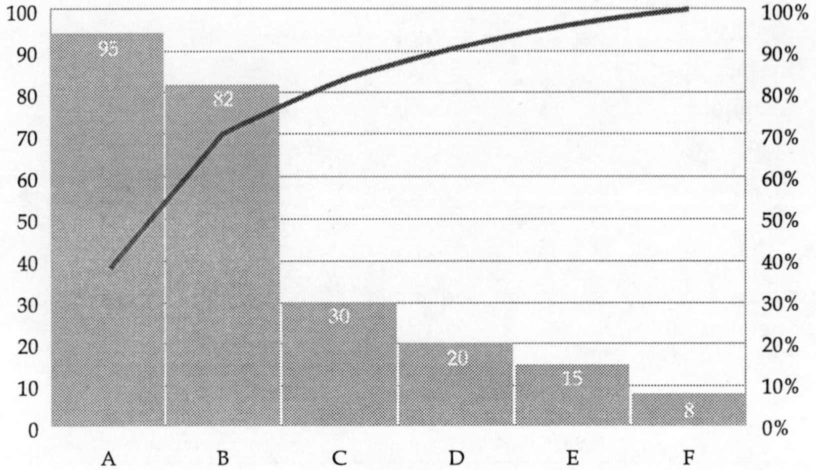

Fig. 2.1

In order to get the maximum benefit from the Pareto analysis, one has to rectify those items which are responsible for the rejection. When that is done one should go for the next lot and observe what the changes are. In the above example, the lot size was 3000. A new lot of 3000 size is to be analyzed. The analysis is given in Table 2.3.

Table 2.3

Sr. No.	*Code*	*Category of defects*	*Frequency*	*Cumulative frequency*	*Cumulative in %*
1.	A	Blowhole	50	50	38.5
2.	B	Incomplete	38	88	67.7
3.	C	Crack	15	103	79.2
4.	D	Surface Scar	12	115	88.5
5.	E	Warpage	8	123	94.6
6.	F	Others	7	130	100
Total			**130**		

While crack and surface scar defects have been reduced but now new defects have come into being and those are blowhole and incomplete. One should work on those and continue till the requirements are satisfied.

Let us introduce "average cost per defect" in Table 2.4 given below.

Table 2.4

Category of defects	*Frequency*	*Average cost per fault*
Blowhole	50	50
Incomplete	38	20

Crack	15	110
Surface Scar	12	50
Warpage	8	300
Others	7	400
Total	**130**	

Table 2.5 is shown where the cost has been taken into account.

Table 2.5

Category of defects	*Total cost*	*Average cost per fault*
Others	2800	50
Blow hole	2500	88
Warpage	2400	300
Crack	1650	400
Incomplete	760	50
Surface scar	600	130
Total	**10710**	

Table 2.6

Category	*Total cost*	*Cumulative total cost*	*Cumulative percentage*
Others	2800	2800	26.1
Blowhole	2500	5300	49.6
Warpage	2400	7700	71.9
Crack	1650	9350	87.3
Incomplete	760	10110	94.4
Surface scar	600	10710	100
Total	**10710**		

(**Source:** Quality Circles, Concept and Techniques by B.R. Dey, Macmillan India Limited, First Published 1988)

Pareto analysis is more useful when one can use in relation to cost. Pareto diagram is a graphical representation that displays data in the order of priority.

2.3.5 Test On Part B (Pareto Analysis)

2.3.5.1 State Which of the Following Statements are True or False

1. Vilfred Pareto was a Nobel Laureate.
2. Pareto worked on the concentration of wealth among the people in Italy.
3. Pereto analysis has many other names.

4. His principle is only useful in the concentration of wealth.
5. This principle is useful to identify the most important causes.

2.3.5.2 *What's Your Choice?*

1. One of the other names of Pareto analysis is
   (a) Principle of 60-40
   (b) This principle is also known as 80-20 principle
   (c) Also known as "trivial few and vital many"
2. Identify where Pareto Analysis is not useful.
   (a) Maintenance of machine
   (b) Inventory control
   (c) Design of products
3. To get the maximum benefit from this technique, it is followed continuously till the requirement is achieved.
   (a) No
   (b) Yes
   (c) Not sure

Answers

True or false

1. (F) 2. (T) 3. (T) 4. (F) 5. (T)

What's your choice?

1. (b) 2. (c) 3. (b)

2.3.6 Problems

Problem 1

An enthusiastic boy of a school has collected the names from the different classes and categorized into seven categories which he identified in alphabets from A to G. The details is shown in Table 2.7.

Table 2.7

Sr. No.	*Names*	*Frequency*
1.	A	11
2.	B	6
3.	C	4
4.	D	48
5.	E	8
6.	F	5
7.	g	18
	Total	**100**

Source: The Home Accident Surveillance System (HASS) data 1999

Compare:

(a) The average number of casualties (of any type) per year of age in each age group.

(b) The data on each type of accident suffered by children in the two age groups.

2.4 ISHIKAWA DIAGRAM

2.4.1 Introduction

The concept of this diagram was initiated by Prof Kaoru Ishikawa in the beginning of 1950s in Japan. On the basis of the creator of this diagram, it is named as Ishikawa diagram. The concept of the diagram is based on the universal law of cause & effect. Because of this it is also called "cause and effect diagram". This diagram was first applied in the year 1953 in Fulsai Iron Works in Japan. The concept is almost seventy years old which indicates its potential even now.

2.4.2 Elements of Ishikawa Diagram

The basic elements of Ishikawa Diagram can be stated as below:

1. The most important element is the "effect".
2. The second element is the various "main causes".
3. The third element is the sub-causes of the each main causes.
4. Depending upon the requirement sub-causes of each sub-causes.

2.4.3 Procedure of Ishikawa Diagram

(a) The Ishikawa diagram should be used by a group instead of individual to get the best result.

(b) The group should adopt "brainstorming technique" to generate ideas.

(c) The identification of "effect" is the prime important.

(d) Identify the main causes which appear to be responsible for effect.

(e) The EFFECT and MAIN CAUSES should be placed in a block.

(f) Against each main cause, identify the sub-cause which is the sub-set of the main cause.

(g) Identify next level of sub-cause.

2.4.4 Why Proper Identification of Effect is Important?

The correct identification will lead to the correct identification of causes. If there is some uneasiness in the body, then it is necessary to know in what exact place it is happening. Wrong identification will lead to wrong cause. If it is headache, then, it will be better if one can say which side and its frequency.

2.4.5 How to Draw Ishikawa Diagram?

1. The "effect" should be put in a block and placed in the right side.
2. Draw an arrow. The head of the arrow should touch the effect block.
3. Inclined lines are to be drawn from the central arrow inclining towards the left.
4. The lines should be both side of the central arrow.
5. Each main cause should be placed on the top of the inclined line and be in the block.
6. Number of inclined lines will be based on the number of main causes
7. Sub-causes should be drawn from the inclined lines parallel to the centre line.

2.4.6 An Example of Ishikawa Diagram

The management has identified that one of their main product's sale has reduced. The market research department comes to the conclusion, comparing the same product in the market that the product cost is high. The problem has been given to a group which has been formed to solve this problem. The group decided to use Ishikawa diagram to solve the problem.

The main effect is identified as "high product cost" The group resorted to brainstorming technique and concluded the following main causes which appear to be responsible for the effect. Those are:

- Design,
- Material
- Labor
- Machines & equipment,
- Overheads.

The group tries to analyze threadbare. Because of this the group identified the sub-causes of the main causes. This is also shown as below:

Design: Old design, collaborator's design

Material: Carbon steel; heavy sections; low yield;

Labor: Unskilled;

Machines & equipment: Scrap World War II machines & equipment; Less productive

Overhead: Administrative; sales

Based on the above data, Ishikawa Diagram is drawn as below:

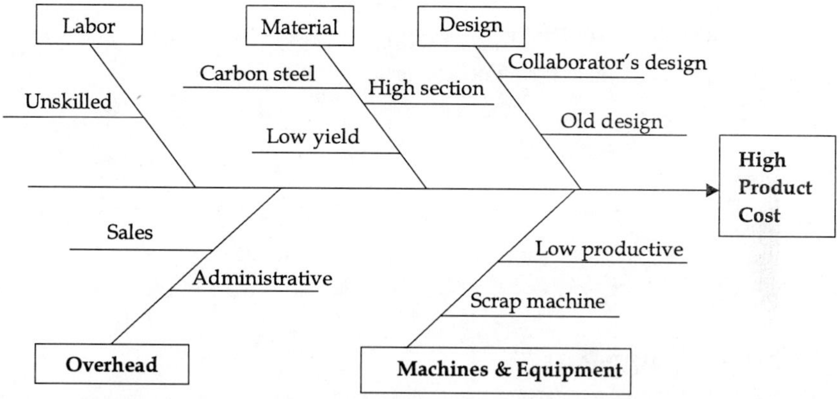

2.4.7 Solution Based on Causes

Based on the causes from the Ishikawa Diagram, the group has come out with the following solutions.

1. They should have their own design where they can modify it to scrap less blanking.
2. Material should be stainless steel.
3. ITI trained labor should be used.
4. CNC machine should replace the old machine.
5. Administrative overheads should be reduced.

2.4.8 Test On Part B (Ishikawa Diagram)

2.4.8.1 True or False

(a) Ishikawa diagram is also called "cause and effect diagram."
(b) It is not a group technique
(c) It was invented by Harold Tufty
(d) "Effect" is to be put on the front side of the arrow in Ishikawa diagram
(e) The diagram was first applied to Fulsai Iron Works in Japan

2.4.8.2 What's Your Choice

1. One of the following is also the name of Ishikawa diagram.
   (a) Fast diagram
   (b) Fishbone diagram
   (c) Normal curve diagram
2. This is one of the seven tools of
   (a) Value engineering
   (b) Theory of constraints
   (c) Quality circle

3. Main causes should be on an --------- line
   (a) Slant line from centre line towards left
   (b) A curve line from the centre line
   (c) Slant line from centre line towards right

Answers

True or false

1. (T) 2. (F) 3. (F) 4. (T) 5. (T)

What's your choice?

1. (b) 2. (c) 3. (a)

2.5 PAYBACK PERIOD

2.5.1 Why Payback is Important?

Any management invests money in new equipment or machinery or anything which is needed for the business in order to get the benefit out of this investment. In the present-day scenario because of continuous technological improvement and innovation, obsolescent of the product, process and system happens regularly. It is because of this reason that the management wants to get its money back at the earliest before it is obsolete.

2.5.2 What is Payback?

The payback period is the time required for the net cash inflows to equal the amount of investment, i.e., the period over which the cash inflows pay-off the investment, the shorter the period, the better is the investment. It is mostly expressed in years. Payback method does not take into account the time value of money. The payback period takes care of the important aspect of risk inherent in investment decisions.

In payback method, while comparing projects, that project which takes less time to back invest money, is being accepted.

2.5.3 How is Payback Period Calculated?

The payback period is calculated by counting the number of years it will take to recover the cash invested in a product, project or system. The formula of payback period is

$$\text{Payback Period} = \frac{\text{Project cost}}{\text{Annual cash inflows}}$$

The cash savings from the new equipment costing \$400,000 is expected to be \$100,000 per year for 10 years. The **payback period** is 4 years (\$400,000 divided by \$100,000 per year.)

2.5.4 Examples

Example 1: The following two investments have been proposed by the management with a yearly benefit for both the investment for five years. On the basis of payback method which investment should be preferable for the management?

Year		*Investment 1 (Rs.)*	*Investment 2 (Rs.)*
	Cash outflow	80,000	1,00,000
1	Cash inflow	15,000	25,000
2	Cash inflow	20,000	25,000
3	Cash inflow	20,000	50,000
4	Cash inflow	25,000	60,000
5	Cash inflow	30,000	75,000

Solution:

Year		*Investment 1 (Rs.)*	*Payback period for investment 1*	*Investment 2 (Rs.)*	*Payback period for investment 2*	*Remark*
	Cash outflow	80,000		1,00,000		Investment 2 is better.
1	Cash inflow	15,000		25,000		
2	Cash inflow	20,000		25,000		
3	Cash inflow	20,000		50,000	3 years	
4	Cash inflow	25,000	4 years	60,000		
5	Cash inflow	30,000		75,000		

Example 2: In order to reduce production bottleneck and increase the production capacity of department X, the team has proposed the following information. Advise which proposal is better.

Details	*Proposal 1*	*Proposal 2*
Cost of M/cs	18,000	36,000
Estimated life (years)	5	6
Increase in revenue (per year)	2000	4000
Estimated saving in wage (per year)	5000	6000
Additional costs incurred to run m/c (per year)	1000	1000
Estimating saving in scrap (per year)	500	800
Estimating saving in direct materials (per year)	500	200
Additional cost of supervision (per year)	1000	1000

Solution:

Details	*Proposal 1 (Rs.)*		*Proposal 2 (Rs.)*	
	Cash outflow		*Cash outflow*	
Cost of machines	18,000		36,000	
Revenue	*Cash outflow*	*Cash inflow*	*Cash outflow*	*Cash inflow*
Increase in revenue (per year)		2000		4000
Estimated saving in wage (per year)		5000		6000
Additional costs incurred to run machine (per year)	1000		1000	
Estimating saving in saving in scrap (per year)		500		800
Estimating saving in direct Materials (per year)		500		200
Additional cost of supervision (per year)	1000		1000	
Total cash out flow	2000		2000	
Total cash in flow		8000		11000
Payback period (years)		18,000/6,000 = 3 years	36,000/9000 = 4 years	
Remarks	Proposal 1 is better			

Example 3: A company has decided to replace one of its machines based on the payback Period. The details of the machines are as below:

Sr. No.	*Details*	*Machine A (Imported)*	*Machine B (Indigenous)*
1.	Investment	\$3,00,000	\$158,120
2.	Cash flow/year	\$1,00,000	-----
3.	Net cash flow/year	-------	\$60,815

Solution:

Sr. No.	*Details*	*Machine A*	*Machine B*
1.	Investment	\$3,00,000	\$158,120
2.	Cash flow/year	\$1,00,000	-----
3.	Net cash flow/year	-------	\$60,815
4.	Payback period	**3 years**	**2.60 years**

Based on **payback period**, machine B is better than machine A.

2.5.5 Merits and Demerits of Payback Method

2.5.5.1 Merits

(a) It is simple to understand.

(b) It does not take complex calculations.

(c) For the management this simple information may help to take some immediate decision.

2.5.5.2 Demerits

(a) It does not take into account the appreciation of money.

(b) Other than simple decision it should not be used for very important decision of the management.

2.5.6 Test on Part B (Payback Period)

2.5.6.1 State Which of the Following Statements are True or False

1. Payback method takes care of value for money.
2. Payback period is expressed in years.
3. The project takes less time to return the invested money is accepted.
4. The payback period does not take care of the important aspect of risk inherent in investment decisions.
5. This method is very simple to understand, hence accepted by the management.

2.5.6.2 What's Your Choice?

1. What is the general unit of indicator of payback period?
   (a) Months (b) Days
   (c) Years
2. Is payback method very simple to understand?
   (a) Yes (b) Not very sure
   (c) No
3. Does it take complex calculation?
   (a) No (b) Yes
   (c) Not sure

Answers

True or false

1. (F) 2. (T) 3. (T) 4. (F) 5. (T)

What's your choice?

1. (c) 2. (a) 3. (b)

2.5.7 Problems

Problem 1

The cash flows of the two projects are given below. Advise the management which project should be accepted.

Year	*Cash flow of project A*	*Cash flow of project B*
0	–1,00,000	–1,00,000
1	50,000	20,000
2	30,000	20,000
3	20,000	20,000
4	10,000	40,000
5	10,000	50,000
6	–	60,000

Solution: The payback period of project A is 3 years whereas Project B is 4 years. Hence, Project A is to be accepted.

Problem 2

The following are the cost details of the Project A and Project B.

Items	*Project A*	*Project B*
Investment	Rs. 60,000	Rs. 1,00,000
Life of the project	6 years	10 years
Amount of cash inflows per year	Rs. 15,000	Rs. 20,000

(a) Based on the payback period, which project is to be accepted?
(b) Will there be any change in the decision if you consider the total life of the project?

Solution:

(a) The payback period of Project A is 4 years and that of Project B is 5 years. Hence, Project A is accepted.
(b) Cash inflow after the payback period of Project B will be Rs, 1,00,000 (5 years x 20,000) is more than that of Project A which is Rs. 30,000 (2 years x 15,000). Hence Project B is better.

Problem 3

In order to reduce production bottlenecks and increase the production capacity of department 'X', of M/s Maloy Engineering Industry, the Planning Department has proposed the following two proposals. Advise employing of payback method which proposal is better.

Details	*Proposal I*	*Proposal II*
Cost of machines (Rs)	24,000	44,000
Estimated life (years)	5	7
Increase in revenue per year	3000	5000
Estimated saving in wage (per year)	5000	6000
Additional tool costs incurred to run m/c (per year)	Nil	1000
Estimating saving in scrap (per year)	500	1000
Estimating saving in direct materials (per year)	500	800
Additional cost of supervision (per year)	1000	800

Solution: Payback period of Proposal; I is 3 years and that of Proposal II is 4 years. Hence, Proposal I is better.

Problem 4

The management of Company X is considering purchasing a new machine to increase the production capacity. The life of the machine is 10 years. Company X wants that the payback period should be 4 years. The details of cash inflow and cash outflow are given below:

Initial cost of the machine is Rs. 37,500

Annual cash inflows are: Rs. 75,000 from sales

Annual cash outflows:

Cost of ingredients: Rs. 45,000

Salary expenses: Rs. 13,500

Maintenance expenses: Rs. 1500

Non-cash expenses:

Depreciation: Rs. 5000

Should the company purchase the new equipment? Use payback method

Solution: Company should purchase because the payback period is 2.5 years less than what is desired.

Problem 5

The management of a company wants to reduce labor cost by installing a new machine. They have identified two types of machine, Machine A and Machine B. The price of machine A and machine B are respectively, Rs. 15,000 and Rs. 18,000. Machine A can reduce annual labor cost by Rs. 3000 and Machine B can reduce annual labor cost by Rs. 4500.

Advise which machine should be purchased. Use payback method.

Solution: Should purchase Machine B.

2.6 RETURN ON INVESTMENT (ROI)

2.6.1 What is Return on Investment (ROI)?

The management on regular basis invest money in different projects, purchasing machines and facilities. It is expected that they will get some benefit out of this investment. This rate of return (ROI) is an indicator for the management to know what extent benefit they are getting.

2.6.2 How to Calculate Return on Investment (ROI)?

The return on investment (ROI) is the ratio of the total benefit accrued divided by the amount of investment made. It may be projected in percentage or to mention the amounts of benefit if one person invests one unit of money.

2.6.3 Where is it Used?

The following table indicates how ROI is used in US Department of Defense.

Value engineering savings for the fiscal year 1986 in the US Department of Defense.

Area	*Invested (US$)*	*Saved (US$)*	*Return on investment (ROI)*
Army	26 million	1.3 billion	50:1
Air Force	39 million	608 million	16:1
Navy	54 million	556 million	10:1
Transportation	701,000	60 million	86:1
Energy	152,000	15.8 million	104:1
Veterans	188,000	8 million	38:1
Interior	100,000	5 million	50:1
GSA	55,000	2.3 million	42:1

(**Source:** Value World, Vol. II, No.1)

2.6.4 Example

Example 1: The details of Project A and Project B are given below:

Items	*Project A (Rs.)*	*Project B (Rs.)*
Total Investment	1,00,000	1,50,000
Net cash inflows during the life of the project	1,00,000	2,70,000
Life of the project (years)	10	15

Which project will be preferable to the management based on ROI?

Solution:

Items	*Product A (Rs.)*	*Product B (Rs.)*
Total investment	1,00,000	1,50,000
Net cash inflows per year	1,00,000/10 = 10,000	2,70,000/15 = 18,000
Rate of return (%)	10,000/1,00,000 = 10	18,000/1,50,000 = 12
Hence,		Project B is better than Project A

Example 2: An automobile company decided to organize a Module I Workshop for their associates for five days to improve the cost effectiveness of their products. The company engaged a value methodology consultant. The expenses of organizing the workshop are as below. Number of associates were ten.

(a) Consultant's expenditure

A.1 Transportation 15,500
A.2 Flight ticket 11,300
A.3 Hotel 28,500
A.4 Fee 1, 00,000
Total **1,55,300**

(b) Expenses on the associates

B.1 Cost per day per person for releasing from their work is 2000 will be **1,00,000** (2000 × 5 × 10)

B.2 Tea, lunch expenses: 500 × 10 × 5 = **25,000**

B.3 Expenses on facilities for five days : 10,000 × 5 = **50,000**

Ten associates were divided into two teams for product improvement of Product A and Product B. The outcome at the end of the workshop was presented to their senior management which is as below.

Items	*Product A (Team A)*	*Product B (Team B)*	*Total savings/year*
Annual volume (No)	5,00,000	25,000	
Existing cost/pc	250	1,700	
Proposed cost/pc	210	1,450	
Saving/pc	40	250	
Improvement (%)	16	14.7	
Annual saving	200,00,00	62,50,000	
Saving from the workshop			262,50,000

The management is interested to know the rate of return of this Module I Workshop. How did you interpret the result?

Solution:

ROI = 262,50,000/3,30,300 = 79.47%

or 79.47:1

Though the improvement of both the teams is more or less same but because of annual volume, Team B's benefit is less than Team A. This indicates the selection of project is also an important factor for Value Methodology.

Example 3: A company has initiated a marketing program which costs \$500,000. It is expected that it will expect to deliver profit \$750,000 in the next four years. Calculate ROI of the program.

Solution: The formula for ROI is as below:

$$\text{ROI} = \frac{\text{Gains} - \text{Investment cost}}{\text{Investment cost}} \times 100 = \frac{750{,}000 - 500{,}000}{500{,}000}$$

$$\text{ROI} = \frac{250{,}000}{500{,}000} \times 100 = 50\%$$

Hence, the ROI of the program is 50%.

Example 4

Year	*Investment in Ashok Industry (A)*		*Investment in Hind Industry (B)*	
	Cash outflow	*Cash inflow*	*Cash outflow*	*Cash inflow*
Now	–\$100		–\$100	
Year 1		\$20		\$70
Year 2		\$30		\$60
Year 3		\$40		\$40
Year 4		\$70		\$30
Year 5		\$80		\$20

1. Based on ROI which will be accepted.
2. Discuss the result critically.

Solution: Notice two features of these cash flow streams that are apparent at once.

1 Ashok Industry has the higher overall net cash flow over five years: Ashok's total net cash flow of \$140 is higher than Hind's figure of \$120.

2. Cash flow timing in the two cases is entirely different. Hind Industry (B) seems to be "front-loaded," which means that more substantial returns come in the earlier years. As a result, the analyst will want to know how the timing differences impact several different cash flow metrics.

Note that the timing differences stand out clearly in a net cash flow graph:

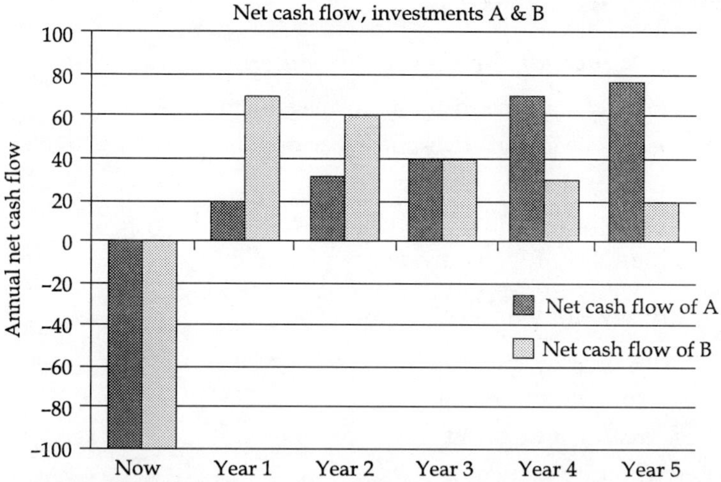

Comparing two cash flow streams: One investment has greater returns early, while the other has substantial returns later.

Example 5: M/s High Tech is interested in producing a new product A. The expected output per year is 40,000 pieces. The variable cost is estimated as $ 30 per piece. The total fixed costs are $800,000. The company has to invest an additional amount of $100,000 in order to produce A. The cost of raising capital is 20%.

Calculate the selling price of A.

Solution:

Sr. No.	*Items regarding product A*	*$*	*Cost per pc ($)*	*Remark*
1.	Variable cost		30	
2.	Fixed cost	800,000	20	800,000/40,000
3.	Total cost		50	
	Desired ROI	(100,000 × 0.20)/ 40,000	5	
	Selling price		55	

2.6.5 Merits and Demerits of ROI

2.6.5.1 Merits

1. It is very easy to calculate.
2. Easy to convince the management.

2.6.5.2 Demerits

1. It cannot be used as the performance of the company.
2. This does not take into account the appreciation of money.

2.6.6 Test on Part B (ROI)

2.6.6.1 State Which of the Following Statements are True or False

1. ROI is not generally used by the management.
2. ROI is the indicator of investment performance.
3. ROI is projected in percentage.
4. It is difficult to convince the management.
5. ROI is the ratio of gains divided by the investment.

2.6.6.2 What's Your Choice?

1. What is ROI?
   (a) Relative ownership of investment
   (b) Resource of investment
   (c) Return on investment
2. Why ROI is so popular a technique?
   (a) Calculation of ROI is easy
   (b) It is not easy to understand
   (c) Convincing the management is difficult
3. How ROI is expressed?
   (a) As a number.
   (b) Usually, expressed as a percentage
   (c) Not as ratio

Answers

True or false

1. (F) 2. (T) 3. (T) 4. (F) 5. (T)

What's your choice?

1. (c) 2. (a) 3. (b)

2.6.7 Problems

Problem 1

You purchased a house at an auction for Rs. 75,000 and spent Rs. 35,000 on renovation. You received Rs.160,000 on the sale of the renovated house. What is the ROI?

Solution: 45%

Problem 2

The details of two projects A & B are given below:

Investment	Project A	1,00,000
	Project B	1,50,000

Life of the project: Project A 10 years
Project B 15 years

Cash inflow during the project life : Project A 1,00,000
Project B 2,70,000

Which project should be selected based on ROI?

Solution: Project B

2.7 DISCOUNT CASH FLOW

2.7.1 Compounding and Discounting

2.7.1.1 Compounding

The time value of money can be explained with the following examples. Let us take that one has deposited Rs. 100.00 in a bank with an interest rate of 10% at simple interest. The following Table 2.10 will indicate how the money changes in the coming five years.

Table 2.10 Money value with simple interest

Year	*Capital amount (Rs.)*	*Interest (Rs)*	*Capital + Interest*
1	100	100 × 0.1	110
2	100	100 × 0.1	120
3	100	100 × 0.1	130
4	100	100 × 0.1	140
5	100	100 × 0.1	150

If the money has been placed with 10% compound interest, the picture will be different as shown below in Table 2.11.

Table 2.11 Money value with compound interest

Year	*Capital amount (Rs.)*	*Interest (Rs)*	*Capital + Interest*
1	100	100 × 0.1	110
2	110	110 × 0.1	121
3	121	121 × 0.1	133.1
4	133.1	133.1 × 0.1	146.41
5	146.41	146.41 × 0.1	163.05

The mathematical formula for compounding is as below:

$$C_n = C_p (1 + r)^n$$

C_N = Compounding value at the end of n years

R = Interest rate

N = Number of years

Problem 2.7.1.1.1

A retired person has deposited in the bank a sum of Rs 1,50,000 at the compound interest of 8%. What he can expect at the end of 6 years?

Solution: In the above problem, C_p is 1,50,000; r = 8%; n = 6 years

Value at the end of 5 years = 1,50,000 $(1 + 0.08)^6$ = 1,50,000 × 1.587 = **2,38,050**

2.7.1.2 Discounting

Discounting is the reverse of compounding. In many cases the management expected some expenditure in future occasion. The management may be interested to know what the present value of that expenditure is. This time one has to use the discounting method. The mathematical formula for discounting will be as below:

$$C_n = C_p / (1 + r)^n$$

C_N = Compounding value at the end of n years

R = Interest rate

N = Number of years

Problem 2.7.1.2.1

The managing director of an organization is expects an expenditure of R, 2,00,000 at the end of 6 years. He is interested to know what the present value is if the interest rate is 10%?

Solution:

In this problem C_p = 2,00,000; r = 10% ; n = 6 years

Present value = 2,00,000 × $1/(1 + 0.1)^6$ = 2,00,000 × 0.565 = 1,13,000

2.7.2 Discounted Cash Flow

Discounted cash flow (DCF) is a valuation method used to estimate the value of an investment based on its future cash flows. DCF analysis finds the present value of expected future cash flows using a discount rate. The present value estimate is then used to evaluate a potential investment.

The formula for *discounted cash flow analysis* is:

$$DCF = CF_1/(1 + r)^1 + CF_2/(1 + r)^2 + CF_3/(1 + r)^3 \ldots + CF_n/(1 + r)^n$$

where:

CF_1 = cash flow in period 1

CF_2 = cash flow in period 2

CF_3 = cash flow in period 3

CF_n = cash flow in period n

r = discount rate (also referred to as the required rate of return)

It is important to note the three most influential components of DCF are time, expected rate of return, and the size of the cash flows of each period. In order to make computation easy a table is given below which is based on year and interest rate.

Present Worth Table

Year	*Interest rate*				
	6%	**8%**	**10%**	**12%**	**14%**
1	0.943	0.953	0.909	0.893	0.877
2	0.890	0.857	0.826	0.797	0,770
3	0.839	0,794	0.753	0.712	0.675
4	0.729	0.735	0.683	0.636	0.592
5	0.747	0.681	0.621	0.567	0.519
6	0.705	0.630	0.565	0.507	0.456

2.7.3 Project Selection Based on DCF Analysis

The management can decide select which project to be selected. This can be illustrated through the examples.

Problem 1

A company has to decide whether it should invest in Project A or Project B. The details of the projects are as below with discounting rate of 10%.

Year	*Project A*		*Project B*	
	Cash outflow	*Cash inflow*	*Cash outflow*	*Cash inflow*
0 Investment	Rs. 30,00,000		Rs. 30,00,000	
1		Rs. 20,00,000		Rs Nil
2		Rs. 40,00,000		Rs Nil
3		Rs. 40,00,000		Rs Nil
4		Rs. 20,00,000		Rs Nil
5		Rs, nil, project closed		Rs. 140,00,000 profit

Solution: The discounted cash flow of both the projects is given below:

Year	*Project A*		*Project B*	
	Actual cash flow	*Discounted cash flow*	*Actual cash flow*	*Discounted cash flow*
1	Rs. 20,00,000	20,00,000 × 0.909 = 18,10,000	Nil	Nil
2	Rs. 40,00,000	40,00,000 × 0.826 = 33,04,000	Nil	Nil
3	Rs. 40,00,000	40,00,000 × 0.751 = 30,44,000	Nil	Nil
4	Rs. 20,00,000	20,00,000 × 0.683 = 13,66,000	Nil	Nil
5	Rs. Nil	Nil × 0.621 = 0	140,00,000	140,00,000 × 0.621 = 86,94,000
	Total	**95,24,000**		**86,94,000**

Project A = Cash inflow – Cash outflow = 95,24,000 – 30,00,000 = 65,24,000

Project B = Cash inflow – Cash outflow = 86,94,000 – 30,00,000 = 56,94,000

Project A is to be selected.

Problem 2

Consider two competing investments in computer equipment. Each calls for an initial cash outlay of $100, and each returns $200 over the next five years making for a net gain of $100. But the timing of the returns is different, as shown in the table below (Case Anjali and Case Swarna), and therefore the present value of each year's gains is different. The sum of each investment's "present values" is called the discounted cash flow (DCF) or "net present value" (NPV). Using a 10% discount rate again, we find:

	Product Anjali	*Product Swarna*
	Net cash flow	*Net cash flow*
Now	– $100.00	– $100.00
Year 1	$60.00	$20.00
Year 2	$60.00	$20.00
Year 3	$40.00	$40.00
Year 4	$20.00	$60.00
Year 5	$20.00	$60.00
Total	**Net CF_A = $100.00**	**Net CF_B = $100.00**

Solution:

Timing	*Product Anjali*		*Product Swarna*	
	Net cash flow	*"Present value"*	*Net cash flow*	*"Present value"*
Now	– $100.00	– $100.00	– $100.00	– $100.00
Year 1	$60.00	$54.54	$20.00	$18.18
Year 2	$60.00	$49.59	$20.00	$16.52
Year 3	$40.00	$30.05	$40.00	$30.05
Year 4	$20.00	$13.70	$60.00	$41.10
Year 5	$20.00	$12.42	$60.00	$37.27
Total	**Net CF_A = $100.00**	**NPV_A = $60.30**	**Net CF_B = $100.00**	**NPV_B = $43.12**

Comparing the two investments, the larger the early returns in Project Anjali lead to a better net present value (NPV) than the later large return in Project Swarna. Note especially the Total line for each present value column in the table. This total is the net present value (NPV) of each cash flow stream. When choosing alternative investments or actions, other things being equal, the one with the higher NPV is a better investment.

2.7.4 Test on Part B (DCF)

2.7.4.1 State Which of the Following Statements are True or False

1. DCF is not a valuation method to estimate the value of an investment based on its future cash flows.
2. Present value estimate is used to evaluate a potential investment.
3. Choosing alternative investments, other things being equal, the one with the higher NPV is a better investment.
4. "Present value" is not the discounted cash flow.
5. Present worth table has three elements: year, rate and investment.

2.7.4.2 What's Your Choice?

1. The full form of DCF is
   (a) Demanded cash flow
   (b) Direct cash flow
   (c) Discounting cash flow
2. What is the relation between compounding and discounting?
   (a) Discounting is the inverse of compounding
   (b) No difference between compounding and discounting
   (c) Discounting is the simple interest
3. Is DCF useful for decision making?
   (a) No
   (b) Yes
   (c) Not sure

Answers

True or false

1. (F) 2. (T) 3. (T) 4. (F) 5. (F)

What's your choice?

1. (c) 2. (a) 3. (b)

2.7.5 Problem

Problem 1

The details of two projects A & B are given below: Interest rate is to be taken as 10%. Please advise which one is better using DCF.

Year	*Project A*		*Project B*	
0	Cash outflow	1,00,000	Cash outflow	70,000
1	Cash inflow	10,000	Cash inflow	5,000
2		15,000		5,000

3		25,000		20,000
4		25,000		25,000
5		20,000		25,000
6		20.,000		30,000

Solution: Project B is better.

2.8 INTERNAL RATE OF RETURN (IRR)

2.8.1 Introduction

The internal rate of return (IRR), also known as the "trial and error method" is to find out the rate which will make the cash outflow, which is at the present value, equal to the present value of the summation of present values of all cost outflows.

In order to do that one has to assume initially some interest rate, find the difference with the cash outflow and the then manipulate the rate with illustrations the rate which will make the present value of cash inflow equal to cash out flow. This is shown by the example given below.

When to compare two or more projects, that project whose IRR is the highest should be chosen. And in any case the IRR should be above the expectations of the management.

Problem 1

Find the IRR of an investment whose initial cash out flow is $ 2, 55,000. The cash inflows for four periods are as below:

Sr. No.	*Years*	*Cash inflows ($)*
1.	1st	65,000
2.	2nd	96,500
3.	3rd	74,100
4.	4th	55,500

Solution:

Year	*Cash inflows*	*Interest rate 5%*	*Present value*	*Interest rate 6%*	*Present value*
1	65,000	0.952	61,880	0.943	61,295
2	96,500	0.907	87,526	0.889	85,789
3	74,100	0.864	64,022	0.839	62,170
4	55,500	0..823	45,677	0.792	43,956
			2,59,105		2,53,210

IRR = 5 + (2,59,105 – 2,55,000)/(2,59,105 – 2,53,210)

= 5 + 4105/5895 = 5 + 0.696

IRR = **5.896%**

2.8.2 Merits

The cash inflows and cash outflows are taken to the present value for comparison.

2.8.3 Demerits

The project life should be taken with care. Any mistake in finding out the project life will lead to wrong decision.

2.8.4 Test on Part B (IRR)

2.8.4.1 State Which of the Following Statements are True or False

1. IRR is also known as trial and error method.
2. IRR is the abbreviation of internal rate ratio.
3. Cash inflows should be converted into the present value.

2.8.4.2 What's Your Choice?

1. **IRR** is a discount **rate** that makes the net present value (NPV) of all cash flows from a particular project equal to zero.
   (a) No
   (b) No
   (c) Yes
2. IRR calculations rely on the same formula as NPV does.
   (a) Yes, there is relation between IRR and NPV
   (b) IIRR has no relation with NPV
   (c) IRR and NPV have no relation
3. Is IRR should be above the expectations of the management?
   (a) No
   (b) Yes
   (c) Not sure

Answers

True or false

1. (T) 2. (F) 3. (T)

What's your choice?

1. (c) 2. (a) 3. (b)

2.8.5 Problem

Problem 1

The cash flow, initial and subsequent three years are given in the following matrix. What is the IRR?

Year	*Cash outflows ($)*	*Cash inflow ($)*
0	1,23,400	
1		38000
2		55800
3		49100

Solution: 7.395%

Problem 2

Find the IRR of an investment having initial cash out flows of Rs. 2,13,000. Cash inflows during the first, second, third and fourth year are expected to be Rs. 65,200, Rs. 96,000, Rs. 73,100 and Rs. 55,400 respectively.

Solution: 14%

2.9 MAKE OR BUY DECISION

2.9.1 Introduction

The make or buy decision is generally used by the manufacturing sector. Initially, the company may manufacture components but over the time when industrial environment changes there is possibility that they can think of buying that component or the assembly instead of manufacturing the component or assembly. Let us study the following situation.

The company is manufacturing a component whose cost structure is as below:

Cost of material per piece: Rs. 370.00

Cost of labour per piece: Rs. 230.00

Cost of overhead per piece: Rs. 250.00

The total cost of the product is Rs. 850.00. If the same component with the same quality is available in the market at Rs. 800.00, then what decision the management is going to take. The decision is obvious. It will be to buy the component instead of manufacturing.

2.9.2 When to Take the Decision

To understand the situation of taking the decision of make or buy, the situation can be narrated as an example.

2.9.2.1 A Case of Complete Assembly

A company which is manufacturing commercial vehicle is manufacturing all the components other than the propriety items. One of the components was leaf spring which is made of different sizes of plates. In order to do that the company was having a separate section in the name of spring shop. The section was purchasing different sizes of plates and manufactured the spring. The process of manufacturing requires machining operations, heat treatment, and finally, assembly. The company has to provide departments like design, planning, quality. Which are dedicated for this component? This was essential because there is no company who is manufacturing this item as per the quality requirement of this company.

But over the years a new company has come up. The company collaborated with them by giving different facilities and they started manufacturing the same quality of spring. As per the given assurance the company started purchasing the item form them and gradually stop their own manufacturing.

The advantage is not only reduction of cost of the component but a lot of advantages like space availability, trained personnel for other new jobs. So, while shifting from manufacturing to buying all points have to be considered.

2.9.2.2 A Case of One Operation

This is a make or buy decision regarding machining operation. The company is the only manufacturer of commercial vehicle. There are many components which are manufactured by the company itself which are made through forging process. For this process each component has to get small blank as per the component which are getting available through cutting from different sizes of billet. In order to do that the company established a cutting section. The cutting section consists of machines like horizontal sawing machine, circular sawing machine, shearing machine and many similar type of cutting machines. The cost of the cutting operation is the part of the product cost.

Now one company is now available which cuts the blank from the billet and supply to the company who needs that. The cost of the machining operation for making a blank is much lower than that of the cost of the company which makes the blank in their cutting section.

The company now decided to **buy** the blank at lower cost from the new company instead of machining in their cutting section. This makes the space available for the company for their future expansion.

2.9.2.3 A Case of New Software

The design department of an organization wants to develop a new product. The new product needs some software. The IT department of the company should make the software. But it does not have the proper resources for making the software. The software is available in the market. Hence, it is correct decision to **buy** than **make** software.

2.9.3 Situation for Using Make or Buy Decision

The factors for making decision whether to make or buy is very important.. In order to do that it is better to clear the following points.

2.9.3.1 Factors Conducive for "Making"

1. The product cost of the item is less than the market price of the same item.
2. The capacity is available with the manufacturer.
3. The item is important and its details cannot be disclosed.
4. The item needs continuous improvement.
5. Reliable suppliers are not available.
6. Political, environmental and social reasons.

2.9.3.2 Factors Conducive for "buying"

1. Market price is cheaper than the item manufactured.
2. Capacity is limited and the manufacturer has to go for new product.
3. Skilled labor is not available
4. Quality of the item is as good as it is being manufactured.
5. Reliable suppliers are available.
6. Needs only small quantity

2.9.4 Test on Part B (Make or Buy)

2.9.4.1 State Which of the Following Statements are True or False

1. Make or buy decision is only applicable for non-manufacturing company.
2. In certain cases even if the making charge is less, one can go for buying.
3. In today's business environment make or buy is an important technique for decision making.
4. Reliable supplier is an important element for taking buying decision.
5. If the quantity is less it is better to go for making.

2.9.4.2 What's Your Choice?

1. In which industry make or buy decision applicable?
   (a) Infrastructure
   (b) Real estate
   (c) Manufacturing
2. One statement in the following is valid for taking buying decision:
   (a) Reliable suppliers are available
   (b) The item needs continuous improvement
   (c) Capacity is available

3. Does skill labor plays a role for taking make or buy decision
   (a) No (b) Yes
   (c) Not sure

Answers

True or false

1. (F) 2. (F) 3. (T) 4. (T) 5. (F)

What's your choice?

1. (c) 2. (a) 3. (b)

2.10 LIFE CYCLE COSTING (LCC)

2.10.1 Introduction

Life cycle cost (LCC), also known as cradle to grave cost, is an important technique in the hands of engineers. When an engineer produces a product, process, project or system, he wants that it should serve properly throughout the life. Not only that the customer should not purchase based on the acquisition cost. There will be ample possibilities that the initial cost may be less but its economic life cost is very high.

This technique will help you to choose the product, process, project or system using this technique.

2.10.2 Methods of Calculating Life Cycle Cost (LCC)

1. Present worth method
2. Annualized method

2.10.2.1 Present Worth Method

2.10.2.1.1 What is Present Worth?

The LCC calculation should be done during the initial phase of the project, whatsoever. All the different types of expenses which will be going to occur will be the future expenses. These expenses will happen in different point of time. The value of money changes with time as shown below:

Appreciation of Money

Year	*Money value at the beginning of the year*	*Money value at the end of the year*
1	100.00	110.00
2	110.00	121.00
3	121.00	133.10
4	133.10	146.41
5	146.41	161.05

Rs 100.00 @ 10% annually compounded

The above table indicates that all the amounts are not the same. Rs 161.05 at the end of 5 years are nothing but 100.00 today. Similarly, 121.00 at the end of 2nd years are 100.00 today. In order to compare alternatives one has to bring all the expenses in one value and that value of money is called present worth. So, Present Worth is the value of all expenses done in various times as of today.

2.10.2.1.2 Present Worth Calculation

In the life cycle cost analysis, if one has to use present worth method, one has calculate all the expenses which are happening at the different points of time into present worth at the starting point of the item. The following example will clarify the point.

Rate of Interest is 10%

Cost elements	*Machine A*		
	Estimated cost	*Present worth factor*	*Present worth*
Cost of M/c	10,00,000		
Cost of installation	12,000		
Acquisitioned cost	10,12,000	1000 (beginning of the year)	10,12,000
Operational cost			
1st year	5,000	0.909	4545
2nd Year	5,000	0.826	4130
3rd year	5,000	0.751	3755
4th year	5,000	0.683	3415
5th year	5,000	0.621	3105
Maintenance cost			
1st year	2,000	0.909	1818
2nd year	2,000	0.826	1652
3rd year	2,000	0.751	1502
4th year	2,000	0.683	1366
5th year	2,000	0.621	1242
Repair & replacement cost			
2nd year	1,000	0.826	826
4th year	2,000	0.683	1366
Salvage cost 5th year	2,00,000	0.621	1,24,200
LCC			**9,16,522**

2.10.2.2 Annualized Method

2.10.2.2.1 Annualized Method Calculation

The above example can be calculated as annualized method as given below. In the annualized method, first calculate the present worth and then multiply with the capital recovery factor, to get the annualized amount.

Rate of Interest is 10%

Cost elements	*Machine A*				
	Estimated cost	*Present worth factor*	*Present worth*	*Capital recovery factor*	*Annualized amount*
Cost of M/c	10,00,000				
Cost of installation	12,000				
Acquisitioned cost	10,12,000	1.000 (beginning of the year)	10,12,000	0.2638	2,66,966
Operational cost					
1st year	5,000	0.909	4545	1.1000	5,000
2nd Year	5,000	0.826	4130	0.5762	2,380
3rd year	5,000	0.751	3755	0.4021	1,510
4th year	5,000	0.683	3415	0.3155	1,077
5th year	5,000	0.621	3105	0.2638	819
Maintenance cost					
1st year	2,000	0.909	1818	1.1000	2000
2nd Year	2,000	0.826	1652	0.5762	952
3rd year	2,000	0.751	1502	0.4021	604
4th year	2,000	0.683	1366	0.3155	431
5th year	2,000	0.621	1242	0.2638	328
Repair & replacement cost					
2nd year	1,000	0.826	826	0.5762	476
4th year	2,000	0.683	1366	0.3155	431
Salvage cost 5th year	2,00,000	0.621	1,24,200	0.2638	32,764
LCC			**9,16,522**		**2,49,303**

2.10.3 List of Tables

For both the methods of calculation for life cycle cost, one requires different tables to bring the cost to the present worth and also to make cost into annualized cost. These tables are:

1. Present worth table
2. Annuity table
3. Capital recovery table

All the above tables have the following elements. The first column is for the years and the other columns are for different interest rates. If one has to find out the factor at the interest rate of 10% in the year 5th year, one has to draw a horizontal line from 5th year and draw a vertical line from the interest rate of 10%, the cutting point will be the factor for that. The factor will be 0.621 This shown in the table below:

Present Worth Table

Year	*Interest rate*				
	6%	8%	10%	12%	14%
1.	0.943	0.953	0.909	0.893	0.877
2.	0.890	0.857	0.826	0.797	0,770
3.	0.839	0,794	0.753	0.712	0.675
4.	0.729	0.735	0.683	0.636	0.592
5.	0.747	0.681	0.621	0.567	0.519
6.	0.705	0.630	0.565	0.507	0.456

These above tables are available in the Appendix.

2.10.4 Costs involved in Life Cycle Cost (LCC)

Any item will have different types of cost during its life. These costs are identified as below:

1. Acquisition cost
2. Operational cost
3. Maintenance cost
4. Repair and replacement cost
5. Salvage cost.

All the five types of costs require elaboration.

Acquisition Cost

This includes along with the cost of the product, the various other costs as below:

(a) Installation cost
(b) Cost of insurance
(c) Cost of registration
(d) Cost of layout, if it is required for the item purchased.

Operational Cost

The operation cost is the expense for running the item. The examples will clarify the point.

1. Fuel consumption (for automobiles)
2. Electric power consumption (for running the machines)
3. Tool cost (for using the machines)

Maintenance Cost

The maintenance costs of the item are those which are required for maintaining the item. The examples are as below:

1. Consumption of engine oil and gear oil, etc. (for automobiles)
2. Lubricants (for machines)
3. Consumption of cleaning materials

Repair and Replacement Cost

1. Repair of floor (building)
2. Change of battery (automobile)
3. Tyre and tube (automobile)
4. Painting –Inner and outside (building)

Salvage Cost

At the end of the product life, if one sells the item he may get the value of one's item.

2.10.5 Assumptions

In LCC calculation one has to deal with various types of cost. These costs do not occur in a particular time but occur throughout the year. In order to minimize the calculation and getting the desired result the following assumptions are made. Since it is a comparison among the different alternatives, following same assumptions for all the alternatives will get the desired result. These assumptions are mentioned below:

1. It is better to use full calendar year for this purpose.
2. Acquisition cost should be taken as happening in the 1st January.
3. Operational cost will be happening, may be every day, but the total operational cost is happening as if on the 31st of the December.
4. Maintenance cost if happening in a particular year it should be considered as happening on 31st December.
5. Repair and replacement cost also think as if it is happening in that year's 31st December.
6. Salvage cost will also happen in the last year's 31st December.
7. The discounting rate will remain same throughout the exercise.

2.10.6 Examples

Example 1: A company has decided to purchase one of two proposals available with them. Company wants to use LCC (present worth method) for decision making. The details of the proposal are shown in below

Description of items	*Machine A*	*Machine B*
Cost of the machine (Rs)	10,00,000	11,00,000
Installation cost (Rs.)	12,000	15,000
Life (Years)	5	5
Operational cost per year (Rs.)	5,000	4,000
Maintenance cost per year (Rs.)	2,000	1,000
Repair and replacement cost (Rs)		
2nd year	1000	
4th year	2,000	1,000
Salvage value (Rs.)	2,00,000	4,00,000
Interest rate (%)	10	10

Solution:

Cost elements	*Machine A*			*Machine B*		
	EST cost	*PWF*	*PW*	*EST, cost*	*PWF*	*PW*
Cost of machine	10,00,000			11,00,000		
Cost of installation	12,000			15,000		
Acquisitioned cost	10,12,000	1.000	10,12,000	11,15,000	1.000	11,15,000
Operational cost						
1st year	5,000	0.909	4545	4,000	0.909	3,636
2nd Year	5,000	0.826	4130	4,000	0.826	3304
3rd year	5,000	0.751	3755	4,000	0.751	3004
4th year	5,000	0.683	3415	4,000	0.683	2732
5th year	5,000	0.621	3105	4,000	0.621	2484
Maintenance cost						
1st year	2,000	0.909	1818	1,000	0.909	909
2nd year	2,000	0.826	1652	1,000	0.826	826
3rd year	2,000	0.751	1502	1,000	0.751	751
4th year	2,000	0.683	1366	1,000	0.683	683
5th year	2,000	0.621	1242	1,000	0.621	621
Repair & replacement cost						
2nd year	1,000	0.826	826			
4th year	2,000	0.683	1366	1,000	0.683	683
Salvage cost	2,00,000	0.621	1,24,200	4,00,000	0.621	2,48,400
LCC			**9,16,522**			**8,96,233**

Since the LCC of Machine B is less, hence it should be accepted.

Example 2: There are two types of refrigerators which have been pinpointed in the initial decision. To make the final decision the company resorted to LCC, using annualized method. The details are given below.

Description of items	*Refrigerator X*	*Refrigerator Y*
Cost of the machine (Rs)	30,000	25,000
Life (years)	5	5
Operational cost per year (Rs.)	400	350
Maintenance cost per year (Rs.)	500	1000
Repair and replacement cost (Rs.)		
3rd year		1,000
4th year	3,000	1,000
Salvage value (Rs.)	5,000	3,000
Interest rate (%)	10	10

Solution:

Cost elements	*Refrigerator X*				*Refrigerator Y*			
	Est. cost	*PWF*	*CR*	*Annualized*	*Est. cost*	*PWF*	*CR*	*Annualized*
Acquisition cost	30,000		0.2638	7914	25,000		0.2638	6595
Operational cost	400			400	350			350
Maintenance cost	500			500	1000			1000
Repair & replacement								
3rd year					1000	0.7513	0.2638	198
4th year	3000	0.6830	0.2638	541	1000	0.6830	0.2638	180
Salvage cost (–)	5000	0.6209	0.2638	819	3000	0.6209	0.2638	491
Life cycle cost (LCC)				**8536**				**7832**

Based on LCC, Refrigerator Y is better.

2.10.7 Limitations of Life Cycle Cost (LCC)

(a) **Life of the project:** Determining the life cycle cost is difficult. While comparing two similar proposals, usually life of both is considered to be same for calculation of life cycle cost. Also component life differs from the whole product life.

(b) **Interest rate:** Interest rate may not be same every year. But we are considering the same interest rate for the entire life of the project.

(c) **Annual expenses:** Annual operation costs are assumed as the expenses occurring at the end of the year while calculating LCC. Intermediate expenses are not taken into consideration.

(d) **Quality/reliability:** We are not considering the quality and reliability of the product while calculating LCC and our focus is only on costs. Here the lowest LCC product may not be a good quality product.

(e) **Aesthetic:** We are not focusing on more one the ESTEEM value of the money spent to improve appearance of the project throughout life while calculating LCC. Hence, in this case the lowest LCC product may fall to add the esteem of the owner even though it provides the use value.

(f) **Comfort & safety:** We are not considering the comfort and safety of the customer while determining LCC. Hence, there is a chance that the lowest LCC product may not provide sufficient comfort/safety resulting in poor value to the customer.

2.10.8 Test on Part B (LCC)

2.10.8.1 State Which of the Following Statements are True or False

1. LCC does not deal with the present worth.
2. There are only two methods of calculating LCC.
3. Salvage cost should be added to get the CC.
4. Though there are limitations of LCC, even then it is utilized to compare the alternatives.
5. We are not considering the quality and reliability of the product while calculating LCC.

2.10.8.2 What's Your Choice?

1. LCC is also known as
   (a) Lower cost consciousness
   (b) Lower cost control
   (c) Cradle-to-grave cost
2. There are two methods; one is present worth, what about the second?
   (a) Annualized
   (b) Average
   (c) Estimated
3. Which does not include the limitation of LCC
   (a) Aesthetic
   (b) Present worth
   (c) Comfort and safety

Answers

True or false

1. (F) 2. (T) 3. (F) 4. (T) 5. (T)

What's your choice?

1. (c) 2. (a) 3. (b)

2.10.9 Problems

Problem 1

The value engineering (VE) department has given the following proposal.

Present System

Assume the design is a fan coil unit (FCU) air conditioning system of 10 tons capacity. The initial cost for the system is Rs. 1800 per ton. The system life is 30 years.

The life of the motors on the FCU's is 12 years and costs Rs. 2000 to replace them. Maintenance cost can be figured at Rs. 1000 per year for the system and utility charges at Rs. 5000 per year.

VE Proposal

The value engineering proposal is to install a variable air volume (VAV) air conditioning system which costs Rs. 1400 per ton. However, to do this requires lowering the suspended ceiling at a cost of Rs. 2000 to accommodate the duct work. It will also cost Rs. 3000 to redesign the system.

However, the system life is 40 years. Yet, every 15 years it costs Rs. 4000 to replace the air handling blowers. Maintenance on this system is Rs. 3000 per year and utility costs are Rs. 4000 per year.

Assume a 10% interest rate on money.

Assume the user needs the building for 20 years.

Applying life cycle cost advice the management which is better.

Solution: VE proposal should be accepted.

Problem 2

Items	*Vehicle I*	*Vehicle II*
Acquisition cost (Rs)	2,75,000	3,25,000
Life (years)	5	5
Operation cost		
Fuel (km/liter)	20	25
Cost of fuel (Rs/liter)	72	72
Annual running (km)	9125	9125
Maintenance cost per year (1st to 3rd)	500(1st); 1000 (2 & 3)	Nil(1st); 500 (2 & 3)
Maintenance cost per year (4th to 5th)	1500(4th); 2000(5th)	1000 (4th & 5th)
Repair & replacement cost		
3rd year (Rs)	12000	
4th year (Rs)		10000
Salvage value (Rs)	25000	95000
Rate of interest (%)	12	12

Based on life cycle cost which vehicle is to be purchased.

Solution: Vehicle II

Problem 3

In order to get a better HVAC System, VE group has suggested two alternatives. The details of the two alternatives along with the present on are given below:

Cost elements	*Present method*	*Alternate I*	*Alternate II*
Initial cost			
Cost of the item	816,000	536,000	738,000
Development cost	5,000	6,000	8,000

Implementation cost	115,000	96,000	152,000
Equipment cost	64,000	64,000	2,000
Annual recurring			
Operation cost per year	30,000	35,000	25,000
Maintenance cost per year	25,000	20,000	16,000
Repair & replacement			
8th year		200,000	
10th year			20,000
16th year	10,000	200,000	
Salvage			
At the end of 20 years	80,000	100,000	75,000

Use the lifespan of 20 years and 10% interest rate, compare the life cycle cost in annualized method and advise the management.

Hint: at the rate of 10

Periodic payment for 20 years is 0.1175

Present worth for 8 years is 0.4665; for 10 years is 0.3855; for 16 years is 0.2176; for 20 years is 0.1486.

3 Target Costing

3.1 ORIGIN OF TARGET COSTING

In the year 1960, the thought of target costing came in Japan. It was observed that in order to compete the western market they have to come out with some new concept. Target costing was started in the manufacturing industry and also in the processing industry. Japan wanted to keep her eyes on Asian market when it was fulfilled by the western goods.

At that time, Japan was facing a lot of problems like shortage of resources and skills to come out with new concept which can be competitive with the western industries.

3.2 DEFINITION OF TARGET COSTING

Target costing is an approach to determine a product's life cycle cost which should be sufficient to develop specified functionality and quality, while ensuring its desired profit. Target cost is finalized by subtracting a desired profit margin from the competitive market price. The target costing reduces the cost as shown below.

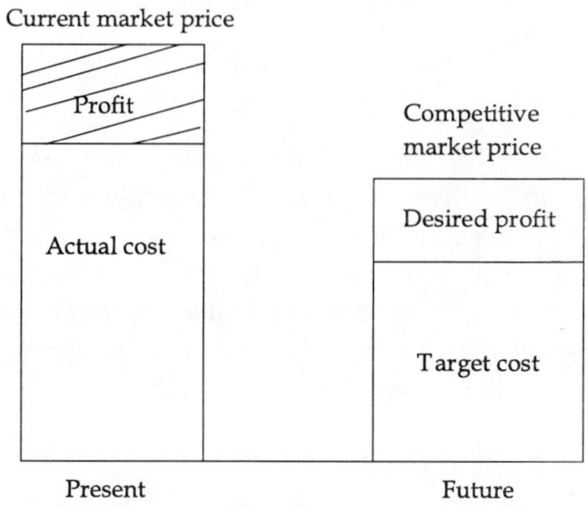

The planned cost reduction is sometimes called "cost drift"

3.3 OBJECTIVES OF TARGET COSTING

The objective of target costing is to make the management proactive in cost planning, cost management and cost reduction. This can be most effectively used at the design and development stage for the new development of the product. The management should monitor the cost in the product life and if there is any deviation, they can rectify those and bring it to the desired level.

Though it is most effectively used for new products, but can also be used for those products which are already launched. Of course, in this case it may not be very effective.

There are mainly three objectives:

1. To lower the cost of the new products so that required profit level can be achieved.
2. The new product should satisfy the quality, delivery and price as expected by the customers.
3. To motivate the employees to be a part of the exercise to maintain the target cost of the new products as well as old products.

3.4 STEPS OF TARGET COSTING

3.4.1 Market Research

The first step is to conduct the market research. They should find the potential areas where they can sell their product. They also have to find out that what are the features the customers are interested in knowing this they can calculate the cost of each feature. It will be easier for them to calculate the target cost of the products having all the features. If it is not viable then they can reduce the features and reduce cost. This information will be very much useful during the full life of the product.

3.4.2 Calculate Price

If the company can decide that they want this much profit margin for the product, then it becomes easy to finalize the price of the product by just adding profit margin decided by the management to the target cost. At this moment they can have two pronged approaches. If any moment they find that price is high they can reduce the profit margin without changing the target cost. Or they can think of reducing the target cost, either by reducing the features or by using some cost reduction technique like value analysis, keeping the profit margin same.

3.4.3 Product Development

Now the step will be product development. Here, two departments are mostly responsible. The first one is design department and the second one is supply chain management, under which the procurement section is. The procurement department will find out the raw materials, components necessary for the product at the best quality and proper delivery not only the raw materials but regarding the components at the lowest possible cost. This department will also responsible to find out whether some of the components can be outsourced.

The Design department will be responsible regarding the materials of the product and the design of the product features. It should be done keeping in mind the quality of raw materials, features which can attract customers with no extra cost. Process of conversion of material should be with zero rejection. The processing machines should be available at the minimum hassle.

3.4.4 Second Look

A multi-disciplined team consists of Value Engineer, Industrial Engineer, Quality Engineer, Marketing and Selling Engineer and the Cost Engineer. The product which has been designed and approved by the management has to pass through this cell. This multi-discipline team will see the product from different angles. Member of marketing and selling engineer will provide the customers 'needs and desires'. These needs and desires will help to make the product more customer friendly. Of course, the team will see that the desire of the customers should not increase the cost. If it is so, the team may come out with balancing strategy.

3.5 TARGET COSTING PROCESS

There is no specific process for target costing. But the following processes may be used with considerable benefit.

1. Establish the target price that the market can bear is essentially for the new product. If one is interested to use the target setting process for the running product, then find out what is the best price in the market which the customer accepts.
2. Establish the profit margin which is expected by the management.
3. Determine the allowable cost which will be accepted by all the stake holders of this product in the company. This is an essential element because then it will become the joint responsibility.
4. In the old product if this allowable cost is less than the current cost of the product, the management has to find out how the current cost can be reduced.

3.6 METHODS OF REDUCING COST

The cost can be reduced by the following ways.

- Simplify the design
- Reduction of direct materials cost.
- Reduction of direct labor cost
- Eliminating waste.

All the ways can also be used at a time. The management can also prioritize the best method, using Pareto diagram. All the methods are explained to have more clarity.

3.6.1 Simplify the Design

Any change of the design of the product is to be done by the design department. While simplify the design the design department should consider the following aspects of the product:

1. Assure convenience
2. Assure dependability
3. Satisfy customer
4. Attract customer

The above aspects are to be elaborated to give proper understanding.

3.6.1.1 Assure Convenience

What is meant by the assure convenience is as below:

(a) Facilitate installation
(b) Instruct customers
(c) Facilitate maintenance
(d) Instruct users
(e) Facilitate service
(f) Indicate availability

3.6.1.2 Assure Dependability

The following should be regarded as an indicator of assure dependability

(a) Facilitate shipment
(b) Resist deterioration
(c) Protect customer
(d) Protect serviceman
(e) Eliminate theft
(f) Protect surroundings

3.6.1.3 Satisfy Customer

(a) Fit surroundings
(b) Ensure adjustment
(c) Minimize selections
(d) Minimize noise
(e) Adequate instruction

(f) Minimize part inventory
(g) Ensure optimum capacity
(h) Optimize quality
(i) Minimize service
(j) Pictorial instruction
(k) Protect environment

3.6.1.4 Attract Customer

Attracting customer is a vital element for the sale of the product. There are a number items which can attract customer, for example, style (product); color combination; space; printing; color; weight; shape of different types like round, square, rectangle etc.; packaging; speed; style (statement); smell

The designer has to a make the combination of various elements to make a unique product for the market.

3.6.2 Reduction of Direct Material Cost

Before going for the reduction of direct materials, let us clarify our thoughts which are considered the direct materials. Can there be any indicator which indicates that this is direct material?

Direct materials have direct relationship with the volume of production. As an example, if one product requires one kg of materials, two products will require two kg of material. If one bearing is required for one product, it will be two bearings for two products. In short, one can say that those items which are considered in "bill of materials" are direct materials.

Direct materials can be categorized in the following way.

3.6.2.1 Raw Materials

The raw materials which have been transformed into the shape of a product by using different process is a direct material. A round bar which is required to be transformed into shaft as per dimensions of the product using machining operations is a direct material. Amount of plastic required to convert into a product through molding is also a direct material. Certain amount of grease which is required for each bearing is also a direct material.

3.6.2.1.1 How to Reduce Cost

The change of design may change from blanking operation to scrappless banking. It will increase the yield which will reduce the cost of the product. Change of process from structural to molding will reduce the cost. Change of material from carbon steel to alloy steel will eliminate painting operation and thereby reduce the cost. Change of material from steel to aluminum may increase the cost but reduce the weight and thereby reduce cost of the product.

3.6.2.2 Bought Out Items

In manufacturing a product, one may use a number of parts which are readily available in the market, like bearings, bolts & nuts, bushes, springs, etc. These items are also considered direct materials, because number of such items is fixed per product. There is a lot of scope for reducing the cost of the products.

3.6.2.2.1 Use Standard Items

One of the advices for reduction of cost is to use standard items instead of specific design items. Use standard bolts and nuts instead of specific ones.

3.6.2.3 Propriety Items

It is difficult to reduce cost of these items. But with the mutual co-operation, the product cost can be reduced.

3.6.2.4 Outsource Items

These items have their own design. So it can be modified by the original equipment manufacturers and also can be done by the outsourced company.

3.6.3 Reduction of Direct Labor Cost

The cost of labor which is connected with the volume of production is conceded as direct labor cost.

The cost can be reduced by different methods as mentioned below:

1. Use high speed machine (CNC) instead of general-purpose machine.
2. One operator works on more than one machine.
3. Qualified operator employed.

3.6.4 Eliminate Waste

This is one of the best methods of reducing the cost of the product. The waste will always associate with the product and its elimination is an important task. The following are the nine types of wastes a company can visualize in everyday operation.

1. Waste caused by work-in-progress
2. Waste caused by defects
3. Waste in equipment
4. Waste in expenses
5. Waste in indirect labor
6. Waste in planning
7. Waste in human resources
8. Waste in operation
9. Waste in startup.

3.7 ADVANTAGES OF TARGET COSTING

(a) It takes care needs and desires of the customers.

(b) It reduces the cost effectively which the market can bear.

(c) It reduces the throughput time.

(d) It reduces the development time of the product.

(e) It is a holistic approach of the management where all, from top to bottom, are involved.

(f) It achieves the best value for money rather than simply reducing the cost of the product.

3.8 REASONS FOR LATE DEVELOPMENT OF TARGET COSTING

Though target costing concept was developed in the first half of 1980, but the importance was felt in the end of the century. One of the main reasons was:

(a) Japanese had developed this concept for their new products. They are also not interested that the world should know about this.

(b) Japanese just-in-time system gained more popularity in the industry, so target costing got the second seat.

3.9 TEST ON TARGET COSTING

3.9.1 State Which of the Following Statements are True or False

1. Target costing was developed in USA.
2. The planned cost reduction is sometimes called as "cost drift".
3. Target cost is finalized by subtracting a desired profit margin from a competitive market price.
4. Japanese just-in-time system is not as popular as target costing.
5. Target costing takes care needs and desires of the customers.

3.9.2 What's Your Choice?

1. Who developed the concept of target costing?

   (a) Russia

   (b) China

   (c) Japan

2. One of the ways can be used to reduce cost.

   (a) Eliminate waste

   (b) Use unskilled operator

   (c) Do not use standard items

3. One of the steps of target costing
   (a) Use aesthetic only
   (b) Market research
   (c) Push for comfort and safety

Answers

True or false

1. (F) 2. (T) 3. (T) 4. (F) 5. (T)

What's your choice?

1. (c) 2. (a) 3. (b)

4 Design to Cost

4.1 INTRODUCTION

Research indicates that the cost of design is only 5% of the product's cost. But the effect of the design is about 60%-70% of the product's cost. The cost of the product because of the irrational design is very difficult to change at later stages. It has also been observed that though design are good technically and functionally but they are not good when the cost is taken into account.

It is clear that there is a demand of proper cost integration into the product at the design stage to get the maximum benefit in the market. This requirement has brought in this concept of design to cost (DTC).

4.2 WHAT IS COST GOVERNANCE?

In the traditional system of business, each department is a separate entity and function as per its own system, of course, with the business objectives of the organization. As an example, where designer is performing his own function of designing the product, planning department responsible for the process after the design is over, will do as per its own system. Finally, the cost of the product is decided based on the above by finance.

But if the organization decides to operate on design to cost system they have to adopt "cost governance". In this system every department will tend to cost and how to reduce that. As an example, during the design stage the planning and cost should be integrated. This will help to optimize the product from all angles. This is because when the product is already launched, it is difficult to modify the design as well as the process to reduce the cost. The following graph (Fig. 4.1) will clarify the point.

Legend:

1. Ability to impact cost and functional capabilities
2. Cost of design changes

4.3 COMPONENTS OF DESIGN TO COST (DTC)

4.3.1 Materials

The percentage of material cost in the product cost varies from 60% to 75%. In auto industries it is approximately 72%. If proper considerations are not made

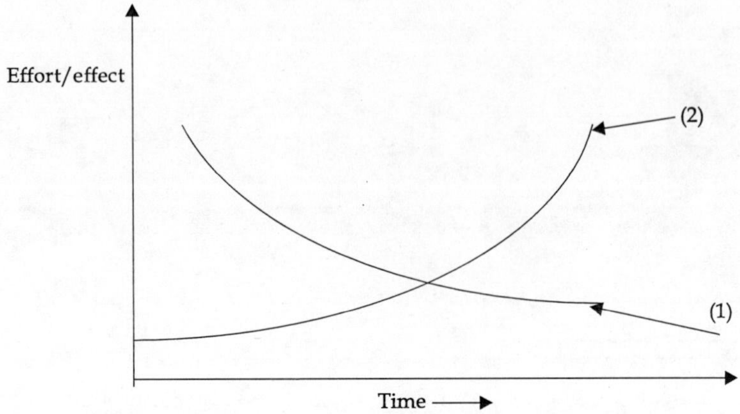

Fig. 4.1

during the design stage, the cost of the product will be forced to increase. If there is a tendency to use higher factor of safety in order to be safe, the weight of the product will invariably increase leading to material cost. It will also have an effect on the processing cost. In the market there are a number of different types of materials available which can replace the earlier material. The designer should come out of the tradition and should utilize the other materials which have higher strength. This may help to reduce the weight of the product and thereby product will be cheaper.

Now, there are various software available. Designer will be able to ensure the quality of the design with different materials.

4.3.2 Manufacturing Process

The cost of manufacturing process is an important element of product cost. The choice of lengthy process creates unnecessary cost of the product. In most of the industries, the manufacturing processes are finalized after the product is finalized by other departments, like planning. Planning department is also responsible to finalize the machines and related equipment.

Hence, it is advisable that during the design process itself, planning department should be involved. This will reduce the cost of the product. Moreover, they can utilize design for manufacturing (DFM) for cost reduction.

4.3.3 Labor

The labor by itself plays an important role in the product's cost. The percentage of labor cost of the product cost varies from product to product. But there are products which are labor oriented. In those cases the labor cost is very high. The designer should be aware about the labor cost per hour for different machines and also the assembly operations. This will help to calculate labor cost at the design stage.

4.3.4 Overhead

Production overheads and administration overheads, are also important ingredients of product's cost. Overhead cost is made up of indirect labor, indirect materials and indirect expenses. As such, judicious utilization is the key for the lower overhead. If this is taken into account at the design stage the cost will be reduced. But minimization of overheads should not affect the quality of product.

4.3.5 Shipment and Logistic

Product design should be such that there should not be any hassle during the transportation of products. This is not only for domestic products but also for export oriented products. Because of design many times the number of pieces to be transported is more, causing unnecessary transportation cost. The heavy product has more effect if exported. The cost of shipping will be high.

The above problems should be addressed during the design stage by involving men from supply chain management.

4.3.6 Side Products

Each product needs some items which are necessary to produce the product as well as for the support of the product. These may be tools, jigs and fixtures, accessories, spares, wrappers for the product, carton for delivery, etc. The cost of these items is also a part of the products cost. This should not be overlooked.

4.3.7 Procurement

The procurement is the activity of purchasing department. The creator of value analysis concept for reduction of product was created by Lawrence Delos Miles. He was working in the purchasing department of GEC (USA). The procurement department can provide some modifications in design by which standard materials can be used to reduce the cost. It is always advisable that procurement division should be included during the design stage.

4.3.8 Maintenance

Some products need maintenance on a regular basis. This will add to the cost. The product designer can reduce this cost in two ways.

1. A more reliable design which requires less maintenance.
2. Maintenance-friendly design for an easier and faster to maintenance.

4.3.9 Quantity

Quantity also casts an important effect on product's cost. It has got an effect on fixed overheads, higher the number less will be the fixed cost. Hence, the product cost can be reduced. During the design stage, the optimum production number can be thought of to reduce product cost.

4.3.10 Design Features

Many a time a feature is added to the product just to attract customer. It should looked into seriously whether it is required or it can be achieved in a better way with less cost.

4.3.11 Finance

The taxes imposed by State or Centre have an effect on cost. This may be customs, VAT etc. It should also be under consideration.

4.4 PROCEDURE FOR DESIGN TO COST (DTC)

In the present business environment "design to cost" has become a requirement for the industry. In order to make the organization a design to cost compliant, it has to follow a system which may be as given below.

4.4.1 Development of Cost Governance Culture

In one of the organizations, I have observed that the cost consciousness was developed among the employees using the following methods.

(a) All the personnel of the organization from worker to the highest level of officers are to attend mandatory program titled "Cost Awareness Program". The group was selected from the same area and provides them the examples from his area along with general ideas where and how they can reduce cost or help to reduce cost. This program boosted up among the employees.

(b) In addition to that every department has to take yearly target in different area for cost reduction. Those targets were the key result areas (KRA) for the department and boiled down to personnel of the department. This KRA were monitored by the top officials every month.

4.4.2 Integration of Design, Process and Cost Engineering

In a conventional way, design, process and cost are different departments. The new product goes through sequentially through these departments. It becomes difficult to change even if it is needed for cost reduction. But if these departments are integrated, then the products will necessarily be full proof.

The integration of otherwise isolated departments will help to remove communication barriers between the three disciplines.

4.4.3 Continuous Monitoring the Market

The organization which will adopt DTC as an initiative should continuously monitor the need, desire and restrictions of the product. This will help to modify the product as and when required before it is too late to keep the organization in the top of the cost effective products.

4.5 CONCLUSION

There is specific tool for design to cost in the market. The cost is an important element of DTC. The cost of various raw materials keeps on varying. The software which can be developed should take care of this. Moreover, the global market situation varies from place to place which has to be kept in mind while designing the software.

1. Shipment and logistics is not a component of DTC (F)
2. Integration of design, process and cost is important (T)

4.6 TEST ON DESIGN TO COST

4.6.1 State Which of the Following Statements are True or False

1. Though the cost of design is only 5% of the product's cost, but the effect of the design is about 60%–70% of the product's cost.
2. Shipment and logistics is not a component of DTC.
3. Integration of design, process and cost is important.

4.6.2 What's Your Choice?

1. Who developed the concept of "design to cost"?
   (a) Japan
   (b) China
   (c) Russia
2. One of the components of "design to cost".
   (a) Sale value
   (b) Quality
   (c) Procurement
3. One of the steps of "design to cost"
   (a) Use aesthetic only
   (b) Market research
   (c) Push for comfort and safety

Answers

True or false

1. (T) 2. (F) 3. (T)

What's your choice?

1. (a) 2. (c) 3. (b)

5 Should Costing/Should Cost Analysis

5.1 HISTORICAL BACKGROUND

Globalization and advances in manufacturing are forcing manufacturers to review and reduce the cost of manufacturing the products. There is a need to identify opportunities for cost reduction through value engineering. ***Should Costing/ Should Cost Analysis*** has the ability to determine a fair and reasonable price for comparative purposes to outsource?

The globalization has made the world a village and any country can take the advantage of less labor cost countries for reducing the cost of the product. The countries which have cheap labor are Russia, Mexico, India, Bangladesh, and Vietnam.

Costing concept was initially thought of by the U.S. Department of Defense. Afterwards, it was included in the Federal Acquisition Regulation (FAR) as a government procurement process. It was adopted by industries in the beginning of 1980. The companies which adopted this technique are Apple and Cisco.

Today, should cost analysis can be applied to many more industries like manufacturing industry, automotive industry, retail and consumer package goods (CPG) industries.

5.2 WHAT IS SHOULD COSTING/SHOULD COST ANALYSIS?

In model based manufacturing and integrated design age, cost is evaluated as one of the design parameters right from preliminary design stage. Designers who typically do not have deep understanding of the manufacturing processes need robust should-cost models that are easy to use and give accurate, directional feedback on design trade-offs.

5.3 MULTI-DISCIPLINARY TEAM APPROACH

Should costing be a robust multi-disciplinary, collaborative and iterative process to optimize design and product cost. It should get detailed cost estimates in order to be competitive. This is possible through Should Costing. The members of the team are:

(a) Engineering department

(b) Procurement department

(c) Supply chain management

5.4 ROLE OF MEMBERS

5.4.1 Engineering Department

Engineering department, during the design stage has clarity about the manufacturing process to produce the product. And in relation to this they are also aware of the type of machines and type of skilled labor required in numbers and skill. So they are responsible to assess the necessary labor.

5.4.2 Procurement Department

Procurement department is responsible for all types of materials. It is not only raw materials but all other materials which will go into the products. It is also responsible for the collection of the expected material cost.

5.4.3 Supply Chain Management

Assigns expected labor rates and profit margins.

Supply chain collects the costs and sums them to determine the should cost.

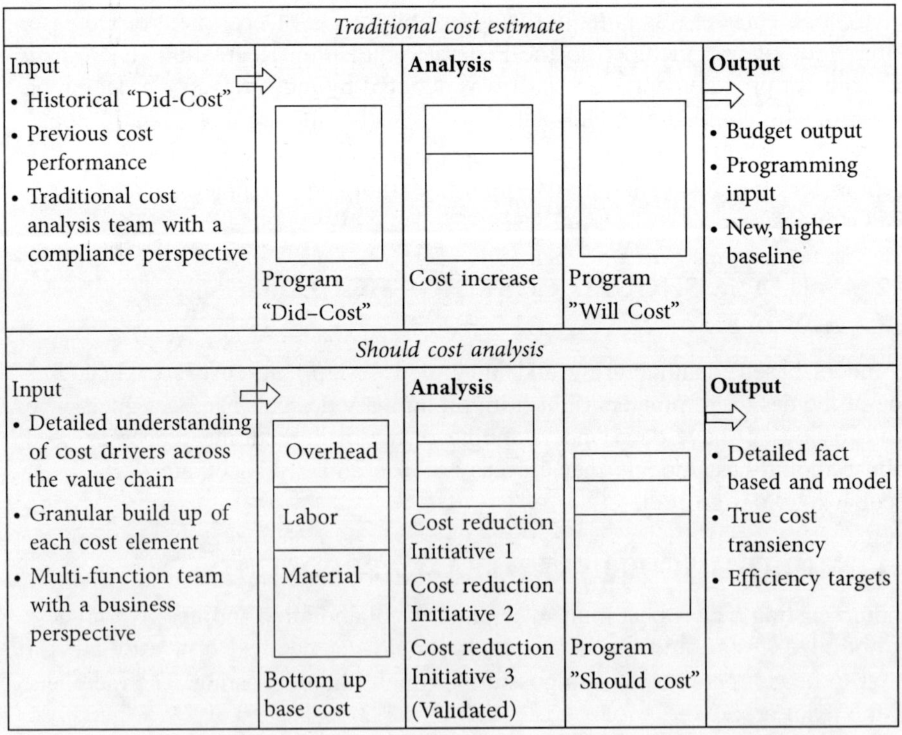

Fig. 5.1 Comparison between traditional cost estimate and should cost analysis

5.5 WHEN TO USE SHOULD COST

Here is a reference for determining which procurement technique is most appropriate depending on the commodities' complexity and number of qualified suppliers.

In the last few years there have been significant advancements in commercially available should-costing tools offering features like automated geometry extraction, automated alternate manufacturing process assessment, advanced materials and machine databases, etc.

Should cost modeling is the process of determining what a product should cost based upon its component raw material costs, manufacturing costs, production overheads, and reasonable profit margins.

5.6 TEST ON SHOULD COSTING

5.6.1 State Which of the Following Statements are True or False

1. Should cost has the ability to determine a fair and reasonable price.
2. It is handled singly.
3. Should costing concept was initially thought of by the U.S. Department of Defense.

5.6.2 What's Your Choice?

1. Cost is evaluated as one of the design parameters right from the preliminary design stage.
   (a) Yes
   (b) No
2. Should cost modeling is not the process of determining what a product should be
   (a) Yes
   (b) No

Answers

True or false

1. (T) 2. (F) 3. (T)

What's your choice?

1. (a) 2. (b)

PART II
Cost Engineering

6 Introduction

Relationship between value and cost; Why knowledge of cost is essential; What is unnecessary cost; Knowledge develops questioning attitude; Importance of managing cost in value engineering.

6.1 RELATIONSHIP BETWEEN VALUE AND COST

The concept of value is not new. Aristotle described it in 365 BC into seven different areas including:

1. Religious value
2. Aesthetic value
3. Political value
4. Judicious value
5. Economic value
6. Social value
7. Ethical value

The concept of value analysis which was founded by Lawrence D. Miles in the fourth decade of the twentieth century, considered only Economic Value and Aesthetic Value. Larry divided these two values into four different types as follows:

1. Cost value
2. Exchange value
3. Esteem value
4. Use value

Miles defined use value as the performance that a product (from now on everything will be called a product whether it is a manufactured product or process or service or system throughout the rest of this book).

Miles defined cost value as the cost of producing products as needed by the customer/user. He defined value in terms of an equation as shown below.

$$\text{Value} = \frac{\text{Performance of required function}}{\text{Cost to acquire those functions}} \tag{1}$$

This mathematical equation indicates that value may be increased or decreased depending upon the relationship of the required functions compared to the cost required to deliver those required functions. This is a clear indication that if one wants to excel in value analysis, one has to learn and should have adequate knowledge of both *cost* and *function*.

6.2 WHY KNOWLEDGE OF COST IS ESSENTIAL?

In many improvement initiatives, the system design is considered untouchable and those initiatives are handicapped due to this logic. However, in the value methodology, the design, as well as the existence of the product, process and system itself must be questioned. The design is not the prerogative of only the designer, but value practitioners can also look into the unnecessary cost embedded within the design.

Similarly, although cost is generally the prerogative of cost accountants, value practitioners should also find out whether there are some unnecessary elements of cost in the products, processes or systems. This is because there are a number of arbitrary assumptions while developing the cost structure of the product, process and system. As an example, overhead distribution to different products is based on a certain approach which may change as the technology changes.

The parent company outsources its products to some other company. They develop their cost structure for negotiations. In the earlier days when there were only conventional machines available, many organizations would allow as much as a ten per cent rejection rate while calculating the cost of a given product. In the advent of NC/CNC machines, those norms had to be modified to one or less than one per cent which would greatly affect the cost of the final product.

Similarly, there have been major improvements in the quality of jigs and fixtures over the years. While it was effective for one lakh (assume) earlier, now it may be double or even more. Similarly, there may be a number of other areas like foundry and forging processes where a lot of changes in technology have taken place in recent years.

Value practitioners must present a number of decisions during the recommendation phase to the management team. The different concepts of life cycle costing, return on investment, payback period have to be developed for obtaining approval by the management in addition to technical viability. To do this job satisfactorily, knowledge of cost is essential.

The above are just some examples where knowledge of cost is a vital and essential requirement for value practitioners. Moreover, these may be potential areas to improve the value of the product, process and system.

6.3 WHAT IS UNNECESSARY COST?

Let me quote Lawrence D. Miles from his epic book "Techniques of Value Analysis and Engineering".

"In achieving better value, there are two separate opportunity areas:

1. The identification of costs as unnecessary
2. The decision making which will eliminate the identified unnecessary cost."

Miles also defined value analysis in terms of unnecessary cost by mentioning that, it is an approach to locate, identify, then remove or eliminate unnecessary cost. He has also defined what is meant by unnecessary cost by saying that costs are considered "necessary" until another means is known to keep the performance factors without them – then they become "unnecessary".

To justify my point I want to quote again from Miles as below.

"Success in marketing a product or service is, to an important extent, depend upon its degree of (1) performance equality or leadership and (2) cost equality or leadership when compared at any time with its competition."

Tolerance is one of the vital areas where unnecessary cost can be associated with. Because of the change of technology and materials and processes, this area needs a special attention for the value methodologists. I have observed some people comparing each dimension of a foundry product where the tolerance can be eliminated because of technology change. It has also been observed that suggestions of dynamic balancing instead of machining operation to make it cost effective.

There is a lot of focus on cost, because as explained earlier it is an element of value. It is also clear that unless one is adequately clear about cost one cannot identify the unnecessary cost.

6.4 KNOWLEDGE DEVELOPS QUESTIONING ATTITUDE

It has been observed that the unnecessary costs embedded ided with the product, service or system are because of wrong belief as well as belief on tradition. It is also known that the development of "questioning attitude" is one of the various traits required to be developed in order to be a good value methodologist. In order to improve the traits mentioned above; knowledge of cost in this case is very essential. This will help to question the criteria used in that arena. The following examples of Miles will help to clarify the matter:

Example 1: "Buy as required – do not accumulate excess inventory".

- The requisition on a one-shot order for screw-machine parts called for 5,000. They cost 10 cents each; total, $500
- The parts could also be made by cold heading; 15,000 was the minimum order. They cost 2 cents each; total $300.
- The above sound general criterion added $200, or 66 per cent extra cost.

Example 2: "Write separate orders when ordering materials from different drawing lists so that incoming packages will be marked with the correct drawing list and accumulation for assembly will be facilitated".

- Small motors were billed to one plant "on the parts list basis" at 28 cents each extra ($2,800 extra for each 10,000 order) because coil assemblies were on one order and rotor assemblies on another.
- The traditional criteria above added $2,800 of extra cost per order.

6.5 IMPORTANCE OF MANAGING COST IN VALUE EQUATION (VE)

Cost has tremendous impact on value. In the present environment cost has to be managed effectively. With reference to our value equation (1), the following Fig. 6.1 narrates the effect of cost.

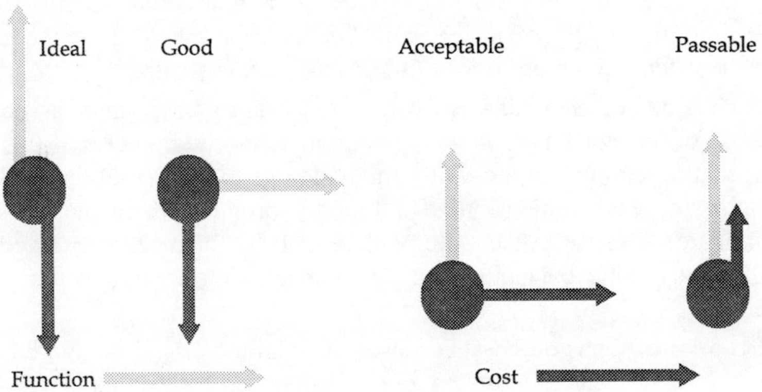

Fig. 6.1 Concepts of cost in relation to value.

Figure 6.1 indicates that the reduction of cost not only improves value but maintenance of cost in the present level can also improve value. Even a slight increase can also improve value if it makes function improved.

The design is now not an untouchable area. Similarly the cost arena also should be probed into by value methodologist. There is a lot of improvement in the cost accounting. If the value methodologist is aware about how the cost has been arrived at, he will be able to advise the cost accountants as they have done to the designers. This will help value methodologists to work in tandem with cost accountants to achieve the cost-effective products.

Objective Type Questions

1. Concept of value is as old as

   (a) 120 BC (b) 120 AD

   (c) 365 BC (d) 365 AD

2. Who had formalized the value?

   (a) Socrates (b) Aristotle

   (c) Pluto (d) Alexander

3. Which value is considered in value methodology?
   (a) Economic value (b) Aesthetic value
   (c) Ethical value (d) Human value
4. Who is the creator of value analysis?
   (a) Aurther E. Mudge (b) John H. Fasal
   (c) Donald E. Parker (d) Lawrence D. Milles
5. One of the values is not economic value?
   (a) Use value (b) Exchange value
   (c) Replacement value (d) Cost value
6. Which cost is the focus of value methodology?
   (a) Production cost (b) Unnecessary cost
   (c) Prime cost (d) Works cost
7. Which attitude is most suitable for value methodologist?
   (a) Criticizing (b) Fault finding
   (c) Questioning (d) Abiding
8. Unnecessary cost is embedded in the product because of ?
   (a) Tradition (b) Good human relation
   (c) Good listening (d) Seeking advice
9. Value is the ratio of?
   (a) Functions by price (b) Price by cost
   (c) Share price by cost (d) Function by cost
10. What is the ideal situation?
   (a) Function same but cost less
   (b) Function more but cost less
   (c) Function more but cost same
   (d) Function increases higher at rate than cost increase.

Answers

1. (c)	2. (b)	3. (a)	4. (d)	5. (c)
6. (b)	7. (c)	8. (a)	9. (d)	10. (b)

7 Elements of Cost

Analysis of total cost; Elements of total cost.

7.1 ANALYSIS OF TOTAL COST

Total cost of a product is an important element for any business. The performance of an organization depends on proper understanding of this total cost. The value of the product depends on that. If it is understood properly one can take necessary steps to work in such a way that unnecessary elements do not enter and vitiate the total cost of the product. That is why, how the total cost is built up should be the starting point in the study of cost. The following indicates the various elements of cost and their nomenclatures.

Table 7.1 Details of total cost

Direct Material	
Direct Labour	
Direct Expenses	
Prime Cost	
Production/Manufacturing Overhead	
Production Cost/Factory Cost	
Administrative Overhead	
Works Cost	
Selling Overhead	
Distribution Overhead	
Total Cost	

In order to understand clearly, each element of total cost is elaborated as follows.

7.1.1 Direct Material

It is essential to have clear knowledge of direct material. Materials having direct relationship with the number of production should be regarded as direct materials.

To have more clarity, if 1 kg of material is required to produce one product, if two kg of materials will be required for manufacturing two products, then it is a direct material. If two bearings are required to manufacture then four bearings will be required for two products since it is a direct material. If this indicator holds good for grease, then grease will also be considered a direct material.

So, to consider a material as direct or not, the above relationship has to be established.

7.1.2 Direct Labour

Direct labour is also similar to direct material. It should have the relation of the production number. If one component takes two hours to complete, four hours will be required for two components.

7.1.3 Direct Expenses

Direct expenses is the third element of total cost. In most of the situations this element is not visible in the total cost. Only on certain cases it will be a part of the total cost. When an enterprise does some expenses particularly for a group of articles, then it will be included as direct expenses. The different situations where these direct expenses are to be considered are:

(a) Hire of special or single purpose tools or equipment for a particular production order or project.
(b) Costs of special layout, designs or drawings.
(c) Maintenance costs of such equipment.

7.1.4 Prime Cost

The summation of all three costs mentioned above is called prime cost. This cost is responsible to convert raw materials into a saleable product. This is the nucleus of the total cost of the product. Any unnecessary cost element ingrained in any of the three elements will be blown up in the subsequent elements. This is because many other cost elements have relation of those elements.

It is an important element which engineers should analyze whenever there is change in the environment because of technological innovations.

7.1.5 Production/Manufacturing Overhead

Before coming to the point, it is necessary to explain what is overhead. Other than those three elements of cost mentioned earlier, there are a lot of expenses which are required to produce the product. But those costs cannot be associated to a particular product as it was with previous elements of cost. These are also in material, labour and expenses and are called:

1. Indirect materials
2. Indirect labours
3. Indirect expenses

The total of three elements above is called overhead. When this happens to be in the production area, it is called Production Overhead/Manufacturing Overhead. It is essential to clarify the different elements.

7.1.6 Indirect Materials

There are a number of materials required to manufacture a product. If the product is manufactured, then it requires machines and to run a machine materials like lubricant oil, grase, spares of the machines and equipment are required. Any of these items costs cannot be ascertained to any particular product and hence considered a part of overhead. Similarly, all the operators require gloves for safety of operation. Operator uses the same for producing a number of different products and cannot be attached with any particular product. Hence, its cost is also a overhead. The following list may help the readers to have better understanding this element of overhead:

1. Consumables like lubricants, grease, and oil for running of machines.
2. Materials, spare parts for repair and maintenance.
3. Tools, dies.
4. Work stationery, hand gloves, soaps.

7.1.7 Indirect Labour

There are some employees who are responsible for production but they may not be for any particular product. As an example, an unskilled worker who is responsible to feed different machines which is manufacturing different products. Their cost is called indirect labour. The foreman and other management personnel are deputed to supervising not only one product but for many more. The cost of them is also considered indirect labour.

7.1.8 Indirect Expenses

There are certain expenses which are also responsible for manufacturing a product. These expenses are known as indirect expenses. As an example, the expenses for rent and freight are one of them. In order to have better idea about these indirect expenses, consider the following examples:

1. Rent and freight
2. Travelling expenses
3. Electricity charges
4. Interest
5. Insurance
6. Canteen expenses
7. Repair and maintenance of plants, building
8. Taxes
9. Medical expenses
10. Store expenses

7.1.9 Production/Manufacturing Cost

The production/manufacturing cost may be defined as the summation of prime cost and production/manufacturing overhead. This may be thought of as the cost of the product at the action area.

7.1.10 Administrative Overhead

Administration is an important aspect for manufacturing a product and it has some costs. This cost has also to be ascertained and is an element of total cost. This cost also includes indirect material, indirect labour and indirect expenses. The following examples provide some input for better understanding.

- **Indirect Material:** Furniture, stationery, postages, books and periodicals, stamps for legal documents, computers, etc.
- **Indirect Labour:** Salary of staff and executives, retirement benefits to executives and staff, perquisite to executives, etc.
- **Indirect Expenses:** Bank charges, hiring charges, telephone expenses, repair and maintenance of office building, rent, tates and taxes, etc.

7.1.11 Works Cost

The summation of production/manufacturing cost and administrative overhead is known as works cost.

7.1.12 Selling Overhead

To produce a good product is not the only objective of any enterprise. The product should be sold to customers also. All expenses required to sell the product is called selling overhead, because it cannot be ascertained to any particular product of the enterprise. The following examples clarify the knowledge regarding selling overhead.

- **Indirect Material:** Furniture, stationery, computers and accessaries, postages, etc.
- **Indirect Labour:** Salary of executives, retirement benefits to executives and staff, perquisites to executives, salary of salesmen and sales office staff, etc.
- **Indirect Expenses:** Advertisement, travelling allowances, commission of salesmen, market research; royalty on sales, etc.

7.1.13 Distribution Overhead

The final element of total cost is distribution overhead. The examples will clarify this element.

- **Indirect Material:** Cartons, paper bag, wooden cases, crates, cylinders, waxed paper, corrugated paper, computer and accessories, etc.

- **Indirect Labour:** Salary of staff and executives, retirement benefits to executives and staff, perquisites to executives, etc.
- **Indirect Expenses:** Carriage outward, warehouse rent, lighting, maintenance of material handling equipment, insurance of finished goods, etc.

7.1.14 Total Cost

The addition of works cost and selling and distribution overhead is the total cost of the product.

Example 1: A small company produces a product which consists of three parts. These parts are supplied by the outside party. The company only assembles the parts to make it a final product. Part A and Part B are assembled in one section and in the final section, the sub-assembly (Part A and Part B) is assembled with Part C to make the final product. The details are given as below:

(a) The cost of Part A is ₹250.00 per 100pcs, Part B costs 180.00 for 100 pcs and Part C costs ₹350.00 per 100 pcs.

(b) The assembly time of Part A and Part B is 300 min per 100 assembled pcs. The labour cost for assembling Part A and Part B is ₹200.00 per day of 8 hrs.

(c) The assembly time for Part A and Part B with Part C is 420 min for 100 assembled pcs. The labour cost for final assembly is ₹240.00 for per day of 8 hrs.

(d) The production overhead is 150% of total labour cost

(e) The administrative overhead is 350% of total labour cost.

(f) The sales and distribution cost is 125% of prime cost.

(g) The profit is 15% of sales price.

Find out the unit cost of the product, its sales price and profit per unit.

Solution:

Item	*Total cost for 100pcs (₹)*	*Cost per unit (₹)*
Material Cost		
Part A	250.00	2.50
Part B	180.00	1.80
Part C	350.00	3.50
Subtotal	780.00	7.80
Labour Cost		
Assembly of Part A and Part B	5*200/8 = 125.00	1.25
Final assembly	7*240/8=210.00	2.10
Subtotal	**335.00**	**3.35**

Contd...

Contd...

Prime Cost	1115.00	11.15
Production overhead	1.50*335.00 = 502.50	5.03
Administrative overhead	3.50*335.00 = 1172.50	11.73
Sales and distribution overhead	1.25*1115.00 = 1393.75	13.94
Total Cost	4183.75	**41.84**
Profit	738.31	**7.38**
Sale Price	4922.06	**49.22**

Ans Product Cost is ₹41.84; Sales Price is 49.22 and Profit is 7.38

Note: Cost + Profit = Price; Cost + 0.15*Sale Price = Sale Price; 4183.75 = 0.85* Sale price

Sale Price = 4183.75/0.85 = 4922.06 and Profit = 0.15*4922.06 = 738.31

Objective Type Questions

1. Pin point the administrative overhead
   (a) Warehouse rent (b) Market research
   (c) Depreciation of machines (d) Materials used
2. One of the items is selling overhead
   (a) Fuel and power
   (b) Royalty on sales
   (c) Maintenance of material handling equipment
   (d) Stamps for legal documents
3. Find out the administrative overhead
   (a) Insurance of finished goods (b) Convertible materials
   (c) Cost of carton (d) None of the above.
4. Find out prime cost element
   (a) Wages of the operators of machines (b) Rents
   (c) Depreciation of machines (d) Cost of cylinders
5. Find out the distribution overhead
   (a) Repairs of building (b) Bank charges
   (c) Warehouse rent (d) Training expenses

Answers

1. (c) 2 (b) 3. (d) 4. (a) 5. (c)

Problems

Problem 7.1

Key	*Items*	*Key*	*Items*
A	Direct material	U	Distribution overhead
B	Rent and rates	V	Direct labour
C	Selling overhead	W	Convertible material
D	Carriage outward	X	Fuel and power
E	Productive labour	Y	Advertisement
F	Production overhead	Z	Administrative overheads

Solution: A to W; B to Z ; C to Y; D to U; E to V ; F to X

Problem 7.2

Give the main groups into which the cost of a manufacturing concern is subdivided. Arrange the following component charges in their respective groups:

Fuel and power	Office salaries
Factory rent	Bank charges
Foreman's wages	Drawing office expenses
Depreciation of plant	Traveller's commission
Productive labour	Convertible material
Repairs of buildings	Bonus
Indirect wages	Indirect material

(Sagar. M.Com., adapted 1955)

Solution:

Items	₹	₹
Convertible material		
Productive labour		
Prime cost		
Fuel and power		
Factory rent		
Foreman's wages		
Depreciation of plant		
Indirect wages		
Works manager's bonus		
Drawing office expenses		

Contd...

Contd...

Indirect material		
Production overhead		
Production cost		
Office salaries		
Bank charges		
Repairs of building		
Administrative overhead		
Works cost		
Travellers' commission		
Sales and distribution overhead		
Total cost		

8 Direct Material Cost

Materials and their importance; distinction of direct material and indirect material; material – a major cost partner; productivity of material; elements of material cost; pricing of material issue; material losses.

8.1 MATERIALS AND THEIR IMPORTANCE

Satisfying the need of the customer is the prime reason for creation of any product, process, service and system. The needed things are not available; they have to be made available. If one wants iron (Fe), it is not available as such. One has to find out something where it is available, and they came to know that it is available in the ore like hematite (70% iron) and magnetite (72%). So to get iron, one has to process either hematite or magnetite. The material which can be converted to get the needed material is called material in cost accounting parlance. So material for iron ore is hematite and magnetite.

Similarly, if one wants to make a box from steel, then steel is the material. If one wants to make from wood, then wood is the material. Similarly, materials will be different if the need of customer is different.

The material plays a vital role in the cost of the product. The material cost in auto industry is 72% of the total cost. Though this varies from industry to industry, but it is always the highest in the cost structure. This is because, engineers have to pay maximum attention to this area.

It is common statement that there is operating loss because of increase of material cost. The use of proper material and its effective utilization can minimize this loss and can also turn into profit. At this point it is not out of place to quote Carlos Fallon which is as below:

"Just knowing what something costs can save five per cent. Improvement in material can save ten per cent."

8.2 DISTINCTION BETWEEN DIRECT MATERIAL AND INDIRECT MATERIAL

The materials are divided into two types. They are direct material and indirect material. The indicator of direct material has already been mentioned in the previous chapter.

An example of water bottle will make the direct material clear. Figure 8.1 is a picture of one litre water bottle. The three components are the bottle, the cap and one litre of water. The process of manufacturing water bottle is rinsed, then one liter of water is filled in, the cap is put on the bottle, finally it is packed for transportation.

Fig. 8.1 Water bottle.

The direct materials are the bottle, the cap and the one litre of processed water. All the three items have relation with total number of production. If one requires hundred water bottles, then one have to arrange hundred bottles, hundred caps and hundred litres of processed water.

Only with the use of direct material the product cannot be produced. The materials which are used to produce a product when cannot be assigned to any particular product or in other words when materials are used for multi-products then we consider that material as indirect material. To produce a product one may require tools. But the same tools may also be used to produce some other product also. So, one considers tools as indirect material. Workers use pairs of hand gloves for operating machine safely. With those pair of hand gloves one produces a number of different products. So, it is an indirect material. Cooling lubricant used to increase the life of the tools. This cooling lubricant is used for not only for one product but for various other products. Hence, again the cooling lubricant is an indirect material. Similarly, there may be a thousand items like these which are considered indirect materials.

8.3 MATERIALS: A MAJOR COST PARTNER

Analyzing Table 8.1, one can find that there are the following items which are the potential constituents/partners of total cost or cost of sales:

- Direct materials
- Direct wages
- Production/factory overhead
- Administrative overhead
- Sales and distribution overhead.

Table 8.1 Data analysis of production statement for the year ending 31st March 2012

	₹	₹
Materials consumed		64,635
Works wages—direct		15,665
Prime cost		80,300
Production overhead		
Works wages–indirect	4,000	
Power	1,050	
Maintenance of plant	3,660	
Rate, light, insurance and heat	900	
Sundry expenses – works	1,400	
Building repairs	90	
Salary for acting as works manager	1,200	
Depreciation of plant & machinery	1,900	
Depreciation of building	720	
Rent of building	1,080	16,000
Production cost		96,300
Administrative overhead		
Office salaries	9,300	
Rate, light, insurance and heat	100	
Sundry expenses – office	2,600	
Building repairs	10	
Salary for acting as office manager	1,000	
Depriciation of building	80	
Rent of building	120	13,300
Works cost		1,09,600
Selling and distribution overhead		
Carriage outwards	2,800	
Travelling expenses	1,200	
Advertising	3,000	
Agents' commission	4,500	
Discount allowed	300	
Bad debts	500	12,300
Cost of sales		1,21,900

An analysis has been done of the data in Table 8.1 which is the annual production statement ending 31.3.2012. The percentage of total cost to the different partners is as given below.

Table 8.2 Percentage of total costs

Potential partners	*Percentage (%)*
Direct material	53
Direct wages	13
Production overhead	13.1
Administrative overhead	10.9
Selling & distribution overhead	10
Total cost	100

The above clearly indicates that material is the major cost partner of the total costs. This varies from industry to industry. In the auto industry it may be as high as seventy-two percentage. Hence, it attracts the attentions of engineers.

8.4 PRODUCTIVITY OF MATERIALS

Productivity is defined as the ratio of output by input. This is not only true for labour productivity but also for material productivity.

In our discussion, it is assumed that materials have been procured as per the design of the product. But even then there is a chance that the material productivity is low. In the case of material, if the required specified input material obtained is equal to output material used, one can say that it is hundred per cent productive. These are the following reasons, which if not properly managed, productivity of material will be affected:

1. Reliable suppliers of material
2. Required amount of purchase
3. Obsolescence to be avoided
4. Proper storing to avoid deterioration of quality
5. Proper handing to avoid damage

8.4.1 Reliable Suppliers of Materials

The materials should not only be of proper specification, but also should be supplied by a reliable supplier. Proper material may be supplied, but timely and adequate supply may be done by some only. The importance of this is because if during the midway the supplier is not in a position to supply adequate material; company may have to purchase the quantity at the higher cost which will affect on the productivity of materials.

8.4.2 Required Amount of Purchase

This is a very important element where there is a chance of unnecessary cost. While purchasing, one should take care of items like safety stock, etc., but many a time purchase is made more than what is required expecting either the continuance of production or expecting unavailability of materials. Now, it should be remembered that the present environment is a dynamic environment. New materials are invented as well as new processes developed or change of design of the product itself requires less material of the same type or change into different material itself. In this situation, more material purchase will become non-movable materials. If it has to be disposed of, it will be at lower price.

8.4.3 Obsolescence to be Avoided

Obsolescence is call of a day. There are certain items which become obsolete at very fast rate. In these types of items one should be very careful to store these items; otherwise these will become non-moving items. Either one has to dispose of at lower value or has to keep it for long.

8.4.4 Proper Storing to Avoid Deterioration of Quality

In order to improve the material productivity, the materials should be kept in such a way that the quality of materials does not deteriorate over the time. The materials which deteriorate faster should be purchased cautiously.

For engineers storing of these types of materials is a potential area for value improvement projects. Any improvement will increase the material productivity.

8.4.5 Proper Handling to Avoid Damage

Handling of materials is as important as storing the materials. Proper method of handling and proper equipment which handle the material are important because it will reduce handling loss. The location of stores should be in such a place so that handling can be optimized. This will help to improve the productivity of materials.

This is also another area where engineers should work upon.

8.5 ELEMENTS OF MATERIALS COST

Engineers work on the total cost of the product. But they should also have knowledgeable to other terms of cost which is known as landed cost. There are various elements of landed cost. Engineers may initiate some projects to improve upon some of those. The elements of landed cost are discussed below.

(a) **Invoice Price:** The price of the product includes the profit element over total costs of the product. The invoice price is finalized after deduction of trade or bulk discounts allowed by the suppliers.

(b) **Duties and Taxes:** Excise duty and sales tax have to be added to arrive at the landed cost. If it is an export item then customs duty has also to be added.

(c) **Containers:** If the containers are not returnable, then the cost of the container has to be added.

This is a green field for the engineers to work upon when there is a demand to reduce the cost. It is one of the areas where engineers have not been exposed extensively.

(d) **Packaging material:** The packaging materials cost also has to be added to get landed cost. The packaging materials are like cartoons, gunny bags, etc. This is also a potential area where unnecessary cost can be reduced or eliminated through improved methods and improved materials.

(e) **Freight:** Freight which one has to pay for getting the material to their stores is known as "carriage inward". It is also to be added for getting the landed cost. This area is also a potential area for engineers.

(f) **Cash discount:** The cash discount is generally allowed by the supplier in order to get their bill cleared at the earliest. One school of thought is to reduce this amount to get the landed cost. Second school of thought is that it should not be deducted but should be treated as a separate entity.

Example 1: M/s Symtex supplied the material, the details of which are as follows.

Invoice price; 5000 units @ ₹5.00 per unit = ₹25,000

Trade discount is 5%

Cash discount is 5% if paid within 15 days.

Delivery charges ₹500.00

Container charges @ ₹100.00 per container of 200 units.

Credit is allowed @ ₹50.00 per container if containers are returned within 90 days.

Calculate the landed cost per unit.

Solution:

Items	*Total cost for 5000 units (₹)*	*Cost per unit (₹)*
Invoice price	**25,000.00**	**5.00**
Less, trade discount 5%	1250.00	0.25
Net price	**23,750.00**	**4.75**
Add, delivery charges	500.00	0.10
	24,250.00	**4.85**
Add, cost of returnable containers		
Charge for containers		
5000/200 = 25*@100 per container = 2,500.00		
Less, credit on return		
25* @50 per container = 1250.00	**1250.00**	**0.25**
Landed cost of material	**25,500.00**	**5.10**

Note: Cash Discount is a managerial policy decision, hence should not be a part of product cost.

8.6 PRICING OF MATERIAL ISSUE

The policy of pricing of material issue has to be decided by the management. There are three methods of pricing the material issue. They are:

1. First-In-First-Out (FIFO)
2. Last-In-First-Out (LIFO)
3. Average method

8.6.1 First-in-First-Out (FIFO)

The material which has been purchased and put into the stores first, has to be issued first. This method of pricing is commonly known as FIFO. This is explained in Table (8.3).

8.6.2 Last-In-First-Out (LIFO)

In the last-in-first-out system, when the issue is made, the last item entered into the stores should be issued first. This in the cost parlance is known as LIFO. With the above example this is explained in Table 8.4.

8.6.3 Average Method

The third method is known as the average method. In this method whenever the material is received, the average is calculated of the new one with all the old ones available. The issue will also make as per the average price evolved. The example has been made in the average method and shown in Table 8.5.

8.7 MATERIAL LOSSES

This is one of the most important areas for the engineers. It has been told earlier that material cost is the highest among all the other elements of costs of the product. By using value methodology, the material loss can be reduced and consequently the cost of the product will be reduced.

The following may be the reasons of material loss:

1. Over-purchasing the material.
2. Lack of proper place to keep the material.
3. Non-maintenance of proper inventory of materials.
4. Not using proper handling equipment.
5. Lack of proper planning.

There are four types of material loss.

1. Wastes
2. Scrap
3. Spoilage
4. Defective

Table 8.3 Stores ledger account (FIFO)

Date	Receipts				Issues				Balance		
	Inv. No.	*Qty. kg*	*Rate per kg*	*Amount*	*Stores req. No.*	*Qty. kg*	*Rate per kg*	*Amount*	*Qty. kg*	*Rate per kg*	*Amount*
			₹	₹			₹	₹		₹	₹
2012											
Dec. 2		200	20	4,000					200	20	4,000
Dec. 5						100	20	2000	100	20	2000
Dec. 10		100	16	1600					100 100	20 16	3600
Dec. 12						100 50	20 16	2800	50	16	800
Dec. 25		100	25	2500					50 100	16 25	3300
Dec. 31						40	16	640	10 100	16 25	2660

Table 8.4 Stores ledger account (LIFO)

Date	Receipts				Issues				Balance		
	Inv. No.	*Qty. kg*	*Rate per kg*	*Amount*	*Stores req. No.*	*Qty. kg*	*Rate per kg*	*Amount*	*Qty. kg*	*Rate per kg*	*Amount*
			₹	₹			₹	₹		₹	₹
2012											
Dec. 2		200	20	4,000					200	20	4,000
Dec. 5						100	20	2000	100	20	2000
Dec. 10		100	16	1600					100 100	20 16	3600
Dec. 12						100 50	16 20	2600	50	20	1000
Dec. 25		100	25	2500					50 100	20 25	3500
Dec. 31						40	25	1000	50 60	20 25	2500

Table 8.5 Stores ledger account (Average)

Date	*Particulars*	*Value of materials in store (₹)*	*Average cost per unit (₹)*	*Value of material issued (₹)*
2012				
Dec. 2	Purchase 200 units @ ₹20.00	4,000		
Dec. 5	Less Issue 100 units @ ₹20.00			2,000
	Balance 100 units @ ₹20.00	2,000		
Dec. 10	Add Purchase 100 units @ ₹16.00	1,600		
	Total 200 units	3,600	18.00	
Dec. 12	Less issue 150 units @ ₹18.00			2,700
	Balance 50 units @ ₹18.00	900		
Dec. 25	Purchase 100 units @ ₹25.00	2500		
	Total 150 units	3400	22.67	
Dec .31	Less Issue 40 units @ ₹22.67			906.80
	Balance 110 units @ ₹22.67	2493.70		

8.7.1 Wastes

Wastes are defined as the basic raw material lost in processing, having no recovery value. Disappearance of quantity or quality of materials during the process such as evaporation, shrinkage and deterioration of quality of certain types of goods, namely, perishable goods are examples of wastes. This area has not yet been touched by value engineers/engineers. New materials or new processes or new methods of handling can reduce these wastes.

8.7.2 Scraps

Scrap is the incidental residue from certain types of manufacturing, usually of small amount and low recoverable value. The examples are like turning generated in turning machines and grinding chips of the grinding machine. If this scrap is generated because of excess machining allowance, then it can be a potential area for improvement.

8.7.3 Spoilage

Spoilage is the loss which arises from the accidental destruction of the value of the work in process, example, for any product which is damaged because of tool damage or sudden power stoppage. This product cannot be processed further and is considered spoilage. If the amount of this is more it can be an item for study.

8.7.4 Defective

Those products which are not acceptable initially, but may be acceptable if additional material, labour and overhead are spent to make it acceptable. As an example, if the product is not manufactured as per drawing because of some missing operation in a particular area, then those products are called defective products. With extra operation and labour, it can be made correct.

At this point it can be mentioned that spoilage item cannot be made correct whereas defective product can be made correct.

Example 2: One parcel containing two important materials are received by a factory and the invoice discloses the following information:

			₹
(I)	Material 1000lbs @ ₹3.00 per lb	=	3,000
(II)	Material 600 lbs @ ₹2.50	=	1,500
	Insurance	=	90
	Sales tax	=	180
	Freight, etc.	=	225

Due to mishandling in the factory's store a loss of 10 lbs of Material I and 9 lbs of Material II is to be allowed. What rate would you adopt for issue of this vital material to jobs? Also calculate the rate, if a provision of 10% is to be made towards probable risk of obsolescence.

(ICWAI, adapted)

Solution:

Part I: Quantity

Particulars	*Material – I*	*Material – II*
Quantity received in lbs.	1000	600
Less: Loss due to mishandling	10	9
Net quantity available for issue	990	591
Less: Provision for probable risk of obsolescence @10%	99	59.1
Effective quantity for issue	891	531.9

Part II: Value

	Material I (₹)	*Material II (₹)*
Value	3000	1500
Insurance @ 2% on value	60	30
Sales Tax @4% on value	120	60
Frieght @ 5% on value	150	75
Total	3330	1665
Rate for charging issues	3330/990 = 3.36 per lb	1665/591 = 2.82 per lb
Revised rate after making provision for risk of obsolescence	3330/891 = 3.74 per lb	1665/531.9 = 3.13 per lb

Objective Type Questions

1. *"Just knowing what something costs can save five per cent. Improvement in material can save ten per cent." Is it true or false?*
   (a) True (b) False
2. A material is direct in one area cannot be indirect material in other area.
   (a) True (b) False
3. One of the reasons responsible for better productivity of material is.
   (a) Required amount of purchase
   (b) Obsolescent is not a factor
   (c) Anybody can be the supplier
   (d) No stress on material handling equipment
4. One of the following is not the material issue system
   (a) First-in-first-out (b) Last-in-first-out
   (c) Last-in-last-out (d) Average method
5. One of the following is not considered material loss
   (a) Wastes (b) High tolerance
   (c) Spoilage (d) Defective
6. If it is an export item then one of the following has also to be added.
   (a) Value Added Tax (b) Profit
   (c) Container cost (d) Customs duty
7. Trade discount is allowed by
   (a) Supplier (b) Customer
   (c) Retailer (d) Wholesaler
8. To add cash discount is the prerogative of the management,
   (a) True (b) False
9. One of the points will affect the productivity of materials
   (a) Reliable suppliers of material
   (b) Required amount of purchase
   (c) Storing to avoid deterioration of quality material is questionable.
   (d) Proper handing to avoid damage
10. Disappearance of quantity or quality of materials during the process such as evaporation falls in which category
    (a) Wastes (b) Scrap
    (c) Spoilage (d) Defective

Answers

1. (a)	2. (b)	3. (a)	4. (c)	5. (b)
6. (d)	7. (a)	8. (a)	9 (c)	10. (a)

Problems

Problem 8.1

Following is the record of receipts and issues of a certain material in the factory during a week:

1983

April 1	Opening balance 50 tonnes @ ₹10 per tonne
April 2	Received 60 tonnes @ ₹10.20 per tonne
April 3	Issued 25 tonnes (stock verification reveals loss of 1 tonne)
April 4	Received back from
Orders	10 tonnes (previous issued @ ₹9.15 per tonne)
April 5 Issued	40 tonnes
April 6 Received	2 tonnes @ ₹10.30 per tonne
April 7. Issued	38 tonnes

Show Store Ledger Accounts by adopting the following two principles.

(i) First in first out

(ii) Last in first out.

(Ranchi B.Com. Hons. Pt. III 1991)

Solution:

(i) Under First in First-out Principles.

Date	*Receipts*			*Issues*			*Balance*		
	Qty in tonnes	*Rate per tonne*	*Value*	*Qty in tonnes*	*Rate per tonne*	*Value*	*Qty in tonnes*	*Rate per tonne*	*Value*
1983		₹	₹		₹	₹		₹	₹
Apr.1							50	10	500
"2	60	10.20	612				50 60	10 10.20	1,112
"3				25	10	250	24* 60	10 10.20	852
"4	10	9.15	91.500				24 60 10	10 10.20 9.15	943.50
"5				24 16	10 10.2	403.20	44 10	10.20 9.15	540.30
"6	2	10.300	20.600				44 10 2	10.20 9.15 10.30	560.90
"7				8	10.200	81.60	36 10 2	10.20 9.15 10.30	479.30

(ii) Under Last in First-out Principles.

Date	Receipts			Issues			Balance		
	Qty in tonnes	Rate per tonne	Value	Qty in tonnes	Rate per tonne	Value	Qty in tonnes	Rate per tonne	Value
1983		₹	₹		₹	₹		₹	₹
Apr.1							50	10	500
"2	60	10.20	612				50 60	10 10.20	1,112
"3				25	10.2	255	50 34*	10.0 10.20	846.80
"4	10	9.15	91.50				50 34 10	10.0 10.20 9.15	938.30
"5				10 30	9.15 10.2	397.50	50 4	10.0 10.2	540.80
"6	2	10.30	20.60				50 4 2	10.00 10.20 10.30	561.40
"7				2 4 2	10.30 10.20 10	81.40	48	10.0	480.00

- Loss on stock verification adjusted on April 3, 1983.

Problem 8.2

A supplier quotes for material A as follows:

Lot price 100 units $1.00 each

500 units $ 0.96 each

1000 units $ 0.90 each

Trade discount 20%; Cash discount 5% in seven days.

Containers charged at $4.00 each; $3.20 credit for return

1 container required for every 150 units.

The purchase decides to buy 600 units.

Transport charges amounting to $13.50 and storage $2.50, were charged by the supplier.

Calculate the purchase price.

(***Source:*** Wheldon's Cost Accountant and Costing Methods)

Solution: **Calculation of Purchase Price**

		Amount($)	*Cost per unit ($)*
Material A:600 units @ $0.96		576.00	0.960
Less Trade discount @ 20%		115.20	0.192
		460.80	0.768
Returnable containers;			
4 containers @ $4.00	16.00		
Less credit on return @$3.20	12.80	3.20	0.005
		464.00	0.773
Transport	13.50		
Storage	2.50	16.00	0.027
		480.00	0.800

Cash discount of $23.04 (5% on 460.80) would be received if payment is made within 7 days, but this would not affect the cost accounts.

9 Direct Labour Cost

What is labour cost; Is labour cost variable? Machine hour rate; How to calculate labour cost per unit; Labour cost vis-à-vis value methodology.

9.1 WHAT IS LABOUR COST?

The direct material as such cannot be converted into a product one desires to bring into existence. Something else is responsible to convert the material into the desired product. This is what is called direct labour.

Direct labour cost is an element of prime cost and is an essential element of total cost. This is measured in terms of hours of the labour's involvement in converting to the desired product. The labour hour is converted to monetary value depending upon which type of labour is involved in the conversion. As an example, the labour hour of one hour involvement of a turner and one hour involvement of grinder will not be same. This is because the monthly payment of grinder will be more than the turner. But if the turning operation is done in NC /CNC machine then the cost may be more than the grinding operation.

Example 1: Calculate the standard labour hour rate for workmen of NC/ CNC operator, grinder and turner from the following data:

Item	*NC/CNC operator*	*Grinder*	*Turner*
Basic pay (PM)	10,000	8,000	5,000
D.A. (PM)	2,000	1,500	1,000
Fringe benefits (PM)	200	150	100
Number of working days per year	300	300	300

Effective hours per day will be 6 hrs.

Solution: **Computation of standard labour hour rate (Figs in ₹)**

Particulars	*NC/ CNC operator*		*Grinder*		*Turner*	
	Monthly	*Annual*	*Monthly*	*Annual*	*Monthly*	*Annual*
Basic pay	10,000	1,20,000	8,000	96,000	5,000	60,000

Contd...

Contd...

D.A.	2,000	24,000	1,500	18,000	1,000	12,000
Fringe Benefits	200	2,400	150	1,800	100	1,200
Total Labour Cost		**1,46,400**		**1,15,800**		**73,200**
No. of working days in a year		300		300		300
Effective hrs per day		6		6		6
Total hrs		1,800		1,800		1,800
Standard Labour rate		81.33		64.33		40.67

Example 2: Calculate the standard labour hour rate for workmen of Grade III from the following data:

Basic pay ₹200 p.m

D.A. ₹150 p.m

Fringe Benefits ₹100 p.m

Leave Rules:

30 days PL with full pay

20 days SL with half pay

Usually, sick leave is fully availed of, what should be the labour cost per hour and if no sick leave is availed of during the year?

(ICWAI – Inter)

Solution: **Computation of Standard Hour Rate**

Particulars	*Monthly amount (₹)*	*Annual amount (₹)*
Basic pay	200	2,400
D.A.	150	1,800
Fringe benefits	100	1,200
Total labour cost	450	5,400
	If sick leave is availed	**If sick leave is not availed**
No. of working days in a year	**300**	**300**
Less: (a) P.L. with full pay;	30	30
(b) Sick leave with half pay 20 days or with full pay	10	
Effective days	**260**	**270**
Total standard hours assuming 6 hrs per day after allowing various allowances	**260 × 6 = 1560**	**270 × 6 = 1620**
Standard labour hour rate	**5400/1560 = ₹3.46**	**620 = ₹3.33**

9.2 IS LABOUR COST VARIABLE?

The labour cost is required to convert the material into a desired product. Same amount of labour in hours will be required for each product; hence it is obvious that it is also variable. The indicator of direct labour is the same as that of direct material. That is, direct labour is also connected with the number of the product, as mentioned. As an example, if to convert direct material to the desired product requires ten hours, then for conversion of two products will be twenty hours.

If the labour cost per unit is ₹ 10, then for producing 10 units, the labour cost will be ₹ 100.This is shown in Fig. 9.1.

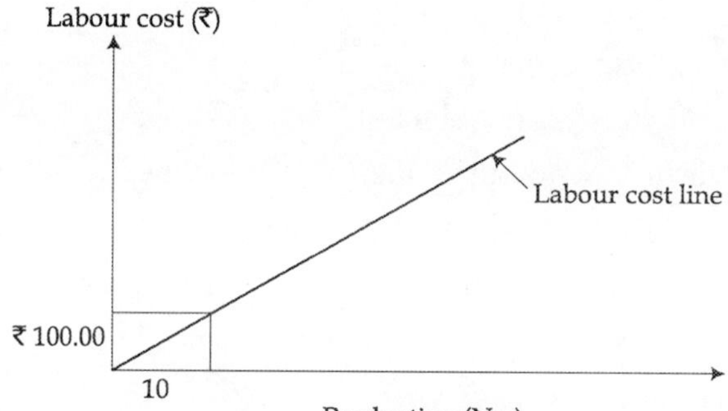

Fig. 9.1 Labour cost.

9.3 MACHINE HOUR RATE

Machine hour rate is another method of calculating the labour cost. The following example will indicate the method.

Example 3: Calculate the machine hour rate for machine 'X' from the following data:

Cost of Machine ₹ 19,200

Estimated scrap value ₹ 1200

Average repairs and maintenance charges per month ₹ 150

Standing charges allocated to machine "X" per month ₹ 50

Effective working life of machine 10,000 hrs

Running time per month 160 hrs

Power used by machine: 5 units per hour at a cost of 19 paise per unit

(C.A. Inter 1969)

Solution: **Computation of Machine Hour Rate of Machine "X"**

Particulars	*Calculation*	₹
Hourly rate of standing charges	(₹ 50/160 hrs)	0.31
Hourly rate of depreciation	(₹ 19,200 – ₹ 1,200)/10,000 hrs	1.80

Contd...

Contd...

Hourly rate of repairs and maintenance charges	(₹150/160 hrs)	0.94
Hourly rate of power used	(5 units × 19 paise)	0.95
Machine hour rate		4.00

Example 4: From the following particulars compute a the machine hour rate for the month of January 1962, to cover overhead expenses of Machine No. 5

Rent of department (space occupied by the machine,
one-fifth of the department): ₹ 780 per annum
Lighting (number of men in the department 12,
Two men engaged on this machine): ₹288 per annum
Insurance etc: ₹36 per annum
Cotton waste, oil etc: ₹60 per annum
Salary of Foreman (one-fourth of the foreman's time
is occupied by this machine and the remainder
equally upon the other two machines): ₹6000 per annum

The cost of the machine is ₹9200 and has an estimated scrap values of ₹200.

It is assumed from past experience:

1. That the machine will work for 1800 hours per annum
2. That it will incur expenditure of ₹1260 in respect of repairs and maintenance over the working life of the machine.
3. That it consumes 5 units of power per hour at the cost of 6 paise per unit.
4. That the working life of the machine will be 18,000 hours.

(Agra. M.Com. 1957; Ranchi, B.Com. 1963A; Delhi, B.Com., 1971)

Solution: **Computation of machine hour rate of machine No. 5**

Items	*Calculation*	*Machine hour rate*
Rent	780/5 × 12 × 150	0,087
Lighting	288 × 2/12 × 12 × 150	0.027
Insurance	36/12 × 150	0.020
Cotton waste, oil, etc.	60/12 × 150	0.033
Salary of foreman	6000/12 × 4 × 150	0,833
Hourly rate of depreciation	(9200-200)/18000	0.50
Hourly rate of repair & maintenance	1260/18000	0.07
Hourly rate of consumption of power	5 units @ 6 paise per unit	0.30
Machine hour rate		1.87

9.4 HOW TO CALCULATE LABOUR COST PER UNIT?

The management is interested to find out the labour cost per unit which is required for decisions making. The method of calculating unit cost of labour is shown through worked out example below.

Example 5: A company is having machine centres which contains different cluster of machines. The finance department of the company has developed machine hour rate of all the machine centres. The machine centre and machine hour rate is as below.

S. No.	*Machine centres*	*Machine hour rate*
1.	Turning machines	1.96
2.	Drilling machines	1.05
3.	Threading machines	1.00
4.	Milling machine	3.25
5.	Grinding machines	4.15

The company is manufacturing shafts of different dimensions and sizes. Recently, the company has received a big order. The planning department has finalized the process of manufacturing of the shaft. The details of the process and the process timing are as below.

Sr. No.	*Process*	*Process timing per piece (in hrs)*
1.	Turning machines	1.75
2.	Threading machines	0.25
3.	Milling machine	0.15
4.	Grinding machines	0.67

You are requested to finalize the labour cost per unit.

Solution: **Computation of Labour Rate per Unit**

Sr. No.	*Process*	*Process timing per piece (in hrs)*	*Machine hour rate*	*Labour cost per unit*
1.	Turning machines	1.75	1.96	**3.43**
2.	Threading machines	0.25	1.00	**0.25**
3.	Milling machine	0.15	3.25	**0.49**
4.	Grinding machines	0.67	4.15	**2.78**
			Total	**6.95**

9.5 LABOUR COST VIS-À-VIS VALUE METHODOLOGY

In order to reduce the cost of the product, the labour cost is an excellent domain for the value methodologists/engineers. Let me quote L.D. Miles, "They didn't show us how to speed up our machines, how to use different types of cutting tools, how to plan some of the parts more effectively. All of this we did ourselves."

The above quotation will indicate a glimpse of the possibility of improvement in the labour cost. Some of the areas where value methodologists/engineers should look into are the following:

1. The processes of converting the raw item into designers' ultimate desired product.
2. The allocation of the standing charges for a particular product.
3. The capacity of utilization of the machines or equipment responsible in the process of changing to desired product, if any.
4. Proper assessment of repairs and maintenance of the machines responsible for conversion.

Objective Type Questions

1. Does direct labour cost vary with the production unit?
   (a) True (b) False
2. Labour cost is not dependent upon on the skill of the operator.
   (a) True (b) False
3. Labour hour rate varies from machine to machine.
   (a) True (b) False
4. One of the following is not included for the calculation of labour rate:
   (a) Basic wage
   (b) Loan for purchasing motorcycle
   (c) Dearness allowance
   (d) Fringe benefit
5. One of the items is called fringe benefit for the labour cost
   (a) Basic wage (b) Dearness allowance
   (c) Profit sharing bonus (d) Loan for purchasing house

Answers

1. (a) 2. (b) 3. (a) 4. (b) 5. (c)

Problems

Problem 9.1

In a factory production is carried with the help of machines and you are supplied with the following information and asked to compute the machine Hour rate:

Original cost of machine ₹20,000

Effective working life of machine 40,000 hrs

Scrap value to be realized ₹1500

Shop rent for the month amounted to ₹488 out of which only 1/4th is to be charged against the machine.

The machine worked for 240 hrs in a month.

Repairs and renewal ₹1000 for the whole life of the machine.

Power used by the machine in a month – 10 units per hour costing 10 paise per unit.

Solution: **Computation of machine hour rate**

	₹
Hourly rate of standing Charges: –1/4th of (₹488/240 hrs)	0.508
Hourly rate of depreciation (₹20,000 – ₹1,500)/40,000 hr	0.463
Hourly rate of repair and renewal (₹1,000/40,000)	0.025
Hourly rate of consumption of power: 10 units @60 paise per unit	1.000
Machine hour rate	**1.996**

Problem 9.2

From the following data, prepare a statement showing the cost per day of 8 hrs of engaging a particular type of labour:

(a) Monthly salary (basic) ₹200.00
(b) Monthly salary (dearness allowance) 5% of basic
(c) Employer's contribution to Provident Fund 8% of salary (Items a & b)
(d) Employer's contribution to State Insurance 2 1/2% of salary (Items a & b)
(e) Prorata expenditure on amenities to labour ₹17.95 per head per month.
(f) No. of working hours in a month 200

(ICWAI-Inter)

Solution: **Computation of labour cost**

	Particular type of labour
No. of working hours in a month	200
Particulars	₹
Monthly salary (basic)	200.00
Monthly salary (dearness allowance) 5% of basic	10.00
Gross wage	210.00
Employer's contribution to providend fund 8%	16.80
Employer's contribution to state insurance $2\frac{1}{2}$%	5.25
Prorata expenditure on amenities to labour	17.95
Total labour cost per month	250.00
Labour cost per hour = 250.00/200	1.25
Labour cost per day of 8 hours	**10.00**

10 Direct Expenses

What is a direct expense; Distinction between direct and indirect expense; When direct expenses to be treated as overhead.

10.1 WHAT IS A DIRECT EXPENSE?

Direct expense is an element of prime cost. It may not have direct relationship with the unit product but has a relationship with a bunch of products. This expense can be attributed to those bunches of products and hence, is called direct expense and associated a part of prime cost. An example will make the point clear. Companies are manufacturing a product on customer request where every tenth piece should be certified by a certifying authority. Because of this stipulation company has to send the tenth piece for certification. The expense involved can be attributed on the products produced. This will have a direct relation.

A few more examples will clarify. They may be like:

(a) Hire of special or single-purpose tools or equipment for a particular production order or project
(b) Costs of special layout, designs or drawings.
(c) Maintenance costs of such equipment.

10.2 DISTINCTION BETWEEN DIRECT AND INDIRECT EXPENSES

While direct expense can be attributed to a unit or of few units, indirect expenses cannot be attributed to a unit or of a few units. It has to be allocated to the unit of products developing certain basis of allocation. As an example, the travelling expenses of a salesman can be allocated to different products may be based on his involvement hours to those products. Because of this indirect expense is not a part of prime cost. It is taken into account in the overhead, which will be dealt in the next chapter.

10.3 WHEN DIRECT EXPENSES TO BE TREATED AS OVERHEAD

The equipment which was earlier used for only particular production order and now used for various other products as well because of method change, then the cost cannot be considered a direct expense. This is because the expense now will be meant for all different products. The involvement of different products may not be same and so the cost incurred will be allocated to different products.

Similar is the case of dies and tools. Dies and tools which were earlier used for some specific products, if extended to various other products because of introduction of new products will also be no longer a direct expense.

The expenses of above types are to be treated as overhead.

Objective Type Questions

1. Direct expenses is not a part of prime cost.
   (a) True (b) False
2. One of the statements is not considered a direct expenses:
   (a) Hire of special or single purpose tools or equipment for a particular production order or project.
   (b) Costs of special layout, designs or drawings.
   (c) Travelling allowance of salesman.
   (d) Maintenance costs of such equipment.

Answers

1. (b) 2. (c)

11 Overhead

Definitions; Classification of overhead; Types of overhead; Allocation, apportionment and absorption of overhead to production.

11.1 DEFINITIONS

There are a number of expenditures which cannot be assigned to any particular product. These expenses are required for the production of products. Since these expenses cannot be directly related to any particular one, they are conceded as a separate identity. This separate identity is called "overhead". All similar type of expenses are collected together and then allocated to different products based on certain logic.

The overhead includes indirect material, indirect labour and indirect expenses. An example will clarify the point. A skill mate is responsible to supply material to three machines manufacturing three different products. His time cannot be allotted to one product; hence it is included in the overhead. Cotton wastes are required for different machines. Since it cannot be connected with any of the product, it is considered an overhead. The travelling expense for the promotion of the products cannot be assigned to any particular product, hence, it is treated as overhead.

11.2 CLASSIFICATION OF OVERHEADS

There are two basic elements in total cost. These are prime cost and the second is overhead. Since the overhead is a measure element of cost, it requires to be analyzed for the different purposes of the business. On the basis of the requirement it is classified into three different categories. Each category is unique by itself and used as and when required. The three categories are as below.

1. Function-wise
2. Element-wise
3. Behavior-wise

11.2.1 Function-wise

The overhead is happening in the different areas of business. The classification has been done accordingly. These are well established and can be enumerated as below.

(a) Manufacturing/Production/Factory overhead
(b) Administrative overhead
(c) Selling overhead
(d) Distribution overhead.

The example of overheads function-wise is shown in Table 11.1.

Table 11.1 Function-wise detailed overheads

Manufacturing	*Administrative*	*Selling*	*Distribution*
Rent of factory	Office salaries	Advertisement	Warehouse Expenses
Depreciation of machines	Office supplies	Salaries of salesmen	Storage expenses
Repairs of plant and machines	Office expenses	Selling agents	Packing and shipping expenses
Power	Postage and courier expenses	Sales office, staff expenses	Carriage outwards
Fuel	Telephone expenses	Commission on sales	Maintenance of transport vans
Factory lighting, heating and air-conditioning	Stationery	Brokerages	Finished stocks
Depreciation of factory building	General administration expenses	Bad debts	Insurance
Sundry consumable stores	Maintenance of factory estate	Market research	
Indirect shop labour	Depreciation of office building		
Tools for general purpose	Director's remunerations		
Insurance on machinery and factory building	Bank charges		
Overtime			
Idle time			
Store-keeping expenses			

The above list is not exhaustive. The overhead allocation function-wise has to be logical and it depends upon situation of its expenditure.

11.2.2 Element-wise

The next category is element-wise. There are three elements in the overhead, whether these are manufacturing or administrative or selling or distribution overheads. These three elements are:

1. Indirect materials
2. Indirect labours
3. Indirect expenses

Table 11.2 indicates the different elements of manufacturing overhead.

Table 11.2 Element-wise manufacturing overhead

Indirect Materials	*Indirect Labour*	*Indirect Expenses*
Stores for repair and maintenance	General indirect labour	Salary of factory staff
Sundry stores for factory use	Idle time	Hospital and dispensary
Small tools for general use	Wages for maintenance workers	Canteen
Fuel	Workmen's compensation	Training expenses
Lubricant oil	Overtime and night shift bonus	Depreciation of plant and machinery and buildings
Deficiencies, loss and deterioration of stores	Holiday pay	Insurance
	Leave pay	Taxes
	Employer's contribution to funds	Rates and rents
	Miscellaneous allowances to labour	

11.2.3 Behaviour-wise

In the behaviour-wise, there are three elements. This way of overhead project is utilized in the various decisions making. These categories are:

1. Fixed overhead
2. Variable overhead
3. Semi-variable or semi-fixed overhead

The semi-variable or semi-fixed has to be divided into variable and fixed overheads.

11.3 TYPES OF OVERHEAD

Behaviour-wise overhead, is very important for the decision making.

11.3.1 Fixed Overhead

Rent is considered a fixed overhead. A company has rented a premise to manufacture product. The rent is ₹10,000 per month. At present the production rate is 1500 units per month. If the production rate changes to 2000 units per month, will there be any change in the rent of the premise? Similarly, if the rate drops to 1000 units per month, then again will there be any change in the rent? In both the cases the rent will not change. This type of overhead which has no relation with production unit is fixed overhead. Some examples of fixed overheads are:

1. Rents and rates
2. Depreciation of machines, plants and buildings
3. Legal charges
4. Insurance
5. Training expenses, etc.

The following example will clarify the effect of fixed overhead on cost per unit. This is shown in Table 11.3.

Table 11.3 Effect of fixed overhead on cost per unit

Items	*April*	*May*	*June*	*July*	*August*	*September*
Production	1430	2000	1000	1430	2000	2500
Rent	10,000	10,000	10,000	10,000	10,000	10,000
Fixed cost per unit	7.00	5.00	10.00	7.00	5.00	4.00

If the fixed cost per unit is analyzed, it will be clear that as the production increases, the fixed cost per unit will decrease. This fixed overhead is an essential item for the value methodologists/engineers because the fixed cost items should be reduced to reduce the cost of the products.

If a curve is drawn between production and the fixed cost per unit, the curve will be as below in Fig. 11.1.

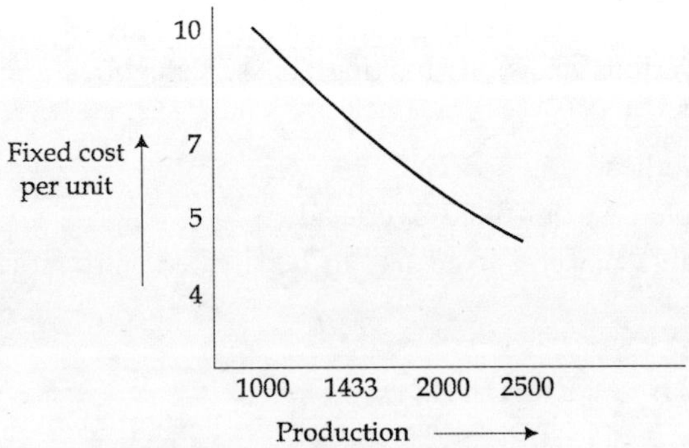

Fig. 11.1 Fixed cost per unit.

11.3.2 Variable Overhead

Overheads which have relation with the production units are called variable overheads. Common examples of variable overheads are:

1. Indirect labours
2. Lubricants

3. Consumable stores
4. General purpose tools
5. Coolants, etc.

The effect of variable overhead is shown in the following Table 11.4.

Table 11.4 Effect of variable overhead in relation to production unit

Items	*April*	*May*	*June*	*July*	*August*	*September*
Production	1430	2000	1000	1430	2000	2500
Variable overhead per unit (₹)	1.50	1.50	1.50	1.50	1.50	1.50
Total variable overhead	2145	3000	1500	2145	3000	3750

Variable overhead is less at the low production. As the production increases the variable cost will increase. The variable overhead per unit line is a straight line and will start from the origin. At zero production the variable overhead will be zero. The graph is shown in Fig. 11.2.

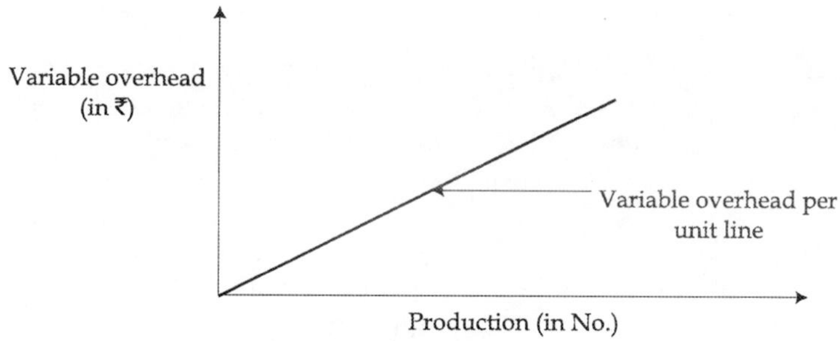

Fig. 11.2 Variable overhead per unit.

11.3.3 Semi-variable or Semi-fixed Overhead

This is the third type of behaviour of overhead. In this type the overhead cost is neither pure fixed nor pure variable. It is mix of both. One type in this is that it is fixed for certain volume of production and changes to different level when the volume changes into the next volume. This type is shown in Fig. 11.3.

The second type which is mixed with both the elements. There are methods which are mentioned below by which the both elements can be determined.

1. By means of equation
2. Graphical method
3. Method of least square
4. High- and low-point method
5. Method of average

The first method is elabourated here.

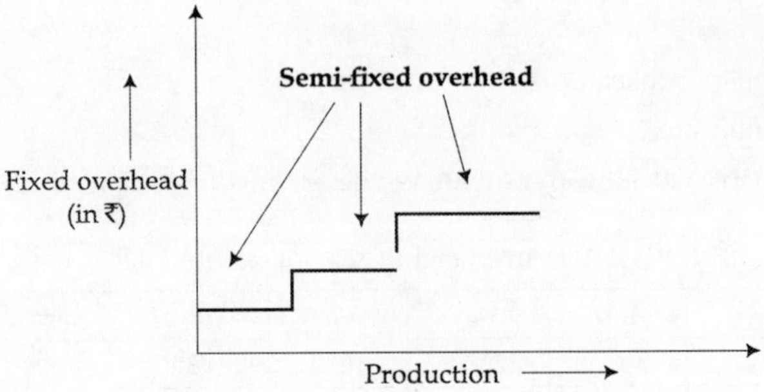

Fig. 11.3 Fixed overhead per unit.

11.3.3.1 By means of equation

The indirect labour expenses for two months are as shown in Table 11.5.

Table 11.5 Month-wise output and indirect labour expenses

Month	*Units of output*	*Indirect labour expenses*
January	500	2,500
July	900	3,700

The assumption is that the pattern will follow the straight line path. The equation of straight line is Y = m.X + C, where Y represents the indirect labour expenses, X is the output and C is the fixed expense and m is the variable expense per unit. This is shown in Fig. 11.4.

Expenses

By putting the above figures in the following equations.

$$2500 = m.\ 500 + C \quad \text{(Equation 1)}$$

$$3700 = m.\ 900 + C \quad \text{(Equation 2)}$$

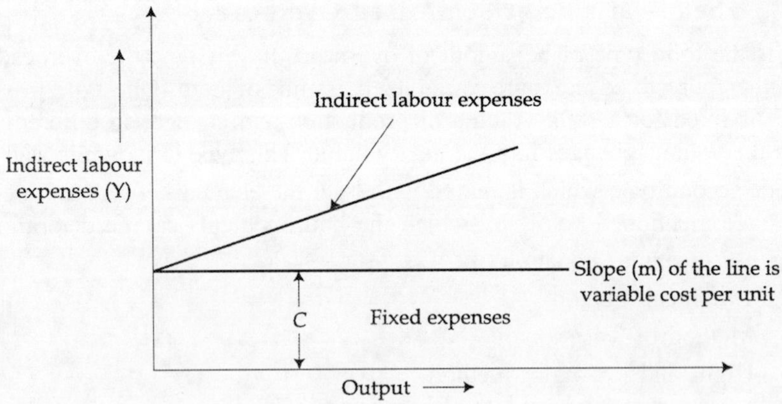

Fig. 11.4 Fixed and variable portion of indirect expenses.

Solving the equations (1) And (2) , we get. M = 3 and C = 1000.

So, the variable cost per unit is ₹3.00 and fixed Cost will be ₹1,000.00.

That means, in the month of January, fixed cost is ₹1000.00 and variable cost is ₹1500.00.

11.4 ALLOCATION, APPORTIONMENT AND ABSORPTION OF OVERHEAD TO PRODUCTION

The product cost is one of the most important aspects of business. Business is viable based on the correct assessment of product cost. The prime cost is directly connected with the product, but the overhead is not directly connected with the product. This portion of expense has to be provided to the product and hence it becomes very crucial. Proper assessment may help to fix the product cost correctly which will help the product to be a cost-effective product. Otherwise, a good product may be a looser product.

Based on the above thought, the overhead is transferred to production in the following methods:

1. Allocation of overhead
2. Apportionment of overhead
3. Absorption of overhead

11.4.1 Allocation of Overhead

The overhead which is associated directly with some production line or cost centre can be directly provided to that production line of cost centre. This method of providing overhead to production or cost centre is called allocation of overhead. This to certain extent can be compared with the elements of prime cost which is having direct relation to production. As an example, if a supervisor is deputed to a particular line only, his salary which is an overhead, will be allocated to production line or cost centre.

It is clear that to have a proper utilization of overhead, attempt to make all the overhead like this. As an example, earlier the consumption of compressed air used by different lines cannot be allocated because there were no proper equipment was available to measure the consumption of compressed air used by different production line. But now a number of equipment have been developed which can measure the total air consumption of the different lines. The same thing is also for power consumption where different equipment can measure properly the consumption of power by different production lines.

This is a green field for value practitioners/engineers who can work in conjunction with cost engineer to come out a value engineered solution.

11.4.2 Apportionment of Overhead

There are a number of items which cannot be projected saying that it is only for this production line or cost centre. Overhead expenses have to be provided to different production lines or cost centres based on certain logical way. The method of providing the overhead to different production lines or cost centres is known as "apportionment of overhead". As an example the company has taken a building on rent where a number of lines have been located. The company is paying yearly rent for that building. Now the question is that the rent cannot be allocated to any production line or cost centre. If one does this there will be unnecessary burden on that production line or cost centre and it may so happen that because of this method a good product will become a non cost-effective product.

Each line may occupy different areas, so the method of apportionment of the rent may be based on "area occupied".

Table 11.6 gives an idea how some overhead has been apportioned.

Table 11.6 Indicators for apportionment of overhead

Sr. No.	*Overhead expenditure*	*Basis of apportionment*
1.	Rent, rate and taxes	Based on floor area
2.	Canteen expenses	Number of employees
3.	Consumable stores	Store requisition
4.	Depreciation of plant, machines and equipment	Capital value
5.	Indirect labour	Number of indirect labour in cost centre
6.	Depreciation of building	Area occupied
7.	Sundry factory expenses	Machine hour or labour hour
8.	Training expenses	Number of employees trained
9.	General purpose tools	Technical estimate
10.	Insurance of plant and machinery	Capital value

In a company along with production areas, there will be a number of service areas. The prime objective of the service areas is to help the production in converting the material into the desired product. These service departments are like, tool room, maintenance department, material handling department, etc. These departments will also be having overhead expenses. Those expenses are also to be apportioned to production. One of the basis of apportionment may be the involvement of service departments to production departments.

Value practitioners/engineers and cost accountants now have to work in tandem to find out the best method of apportionment of overhead. The reasons are that there are a number of new types of expenses are coming up, which have to be apportioned. At the same time a number of inventions are being done to make the expenditure more focussed towards production line or cost centre.

If basis are not available, value practitioners/engineers can come up with new solutions with the help of function analysis, as they have done in various designs.

11.4.3 Absorption of Overhead

With the method of allocations and apportionment, the overheads have been put to different production lines or cost centres. But this is not the end of the process. The cost has to be associated to the products. The method of doing that is known as absorption of overhead. This process is very important because if the method is not proper, it will overload or underload the product which will make the product a loss-making or profit-making product.

The following are the methods used for absorption of overhead:

1. Percentage of direct material cost
2. Percentage of direct labour cost
3. Percentage of prime cost
4. Labour hour rate
5. Machine hour rate
6. Rate per unit of output

Table 11.7 is a comparison of different methods. (see p. 154)

Example 1: The following information relates to the activities of a production department of M/S Hi-tech Ltd. for the year 1980-81.

Material consumed	₹85,000
Direct wages	₹32,000
Labour hours worked	9,500
Machine hours	3,800
Overhead chargeable to the department	₹35,500

On a job executed in that department, the following costs were incurred:

Material consumed	₹6, 000
Direct wages	₹2,100
Labour hour worked	625
Machine hours worked	250

Calculate the overheads chargeable to the job by three commonly used methods.

Solution: Overhead chargeable to the different methods.

(a) On the basis of direct wage cost: (₹35,500/₹32,000) × 2,100 = ₹2329.69

(b) On the basis of labour hours: (₹ 35,500/9,500) × 625 = 2335.53

(c) On the basis of machine hours: (₹35,500/3,800) × 250 = 2335.53

Example 2: Gear India Ltd. is manufacturing three types of gears. They are spur gear, helical gears and bevel gears which are manufactured in three different production departments. In order to serve those departments it has one tool room.

Table 11.7 Comparisons of different methods

Method	*Overhead cost*	*Different basis*						*Indicator (%)/ ₹*
		Direct material cost	*Direct labour cost*	*Prime cost*	*Labour hour*	*Machine hour*	*Output*	
Percentage of direct material cost	1,00,000	2,00,000						50%
Percentage of direct labour cost	1,00,000		50,000					200%
Percentage of prime cost	1,00,000			2,50,000				40%
Labour hour rate	1,00,000				25,000			₹4.00
Machine Hour rate	1,00,000					10,000		₹10.00
Rate per unit of output	1,00,000						50,000	₹2.00

The following figures are available for the month of March which is of 25 working days of 16 hrs each day. All departments work all these days with full attendance.

Expenses	*Total*	*Tool room*	*Spur gear*	*Helical gear*	*Bevel gear*
	₹	₹	₹	₹	₹
Power and lighting	11,000	2,400	2,000	3,000	3,600
Supervisor's salary	20,000	–	–	–	–
Rent	5,000	–	–	–	–
Welfare	6,000	–	–	–	–
Others	12,000	2000	2000	4000	4000
	54,000	–	–	–	–
Supervisor's salary		20%	30%	30%	20%
Number of workers		10	30	40	20
Floor area in sq.ft.		500	600	800	600
Service rendered by tool room to production departments		–	50%	30%	20%

Calculate the labour hour rate of each of the department, spur gear, helical gear and bevel gear.

Solution: **Summary of Distribution of Overhead**

Item of expenses	*Basis of allocation/ apportionment*	*Total*	*Tool room*	*Spur gear*	*Helical gear*	*Bevel gear*
Power and lighting	Direct	11,000	2,400	2,000	3,000	3,600
Supervisor's salary	Percentage	20,000	4,000	6,000	6,000	4,000
Rent	Floor area (5:6:8:6)	5,000	1,000	1,200	1,600	1,200
Welfare	No. of workers (1:3:4:2)	6,000	600	1,800	2,400	1,200
Others	Direct	12,000	2,000	2,000	4,000	4,000
Total		54,000	10,000	13,000	17,000	14,000
Apportionment of tool room to production department	Ratio (5:3:2)		(–10,000)	5,000	3,000	2,000
Total overhead		54,000		18,000	20,000	16,000

Calculation of labour hour rate

Items	*Spur gear*	*Helical gear*	*Bevel gear*
Working hours in a month (25 × 16)	400	400	400
Total labour hours in a month	(30 × 400) = 12,000	(40 × 400) = 16000	(20 × 400) = 8000
Labour hour rate	₹18,000/12,000	₹20,000/16,000	₹16,000/8000
	₹1.50 per hour	₹1.25 per hour	₹2.00 per hour

Objective Type Questions

1. One of the following items is not a part of overhead.
   (a) Rent and rates (b) Depreciation of machines
   (c) Raw material consumption (d) Interest
2. One of the methods of transferring overhead to production is,
   (a) Calculation of overhead
   (b) Apportionment of overhead
   (c) Reduction of overhead
3. One of the methods is not the correct method to find out semi-variable overhead.
   (a) By means of equation (b) Graphical method
   (c) Method of least square (d) By means of fixed table.
4. Overhead is classified into three different categories,
   (a) True (b) False
5. One of the methods is not used for absorption of overhead.
   (a) Percentage of direct material cost
   (b) Percentage of direct labour cost
   (c) Percentage of total cost
   (d) Labour hour rate
6. One of the items is not a variable overhead
   (a) Indirect labours (b) Lubricants
   (c) Consumable stores (d) Direct labours
7. There are three ways of categorizing the overhead.
   (a) False (b) True
8. One of the statement is called overhead.
   (a) Direct material plus direct labour plus direct expenses
   (b) Direct material plus indirect labour plus indirect expenses
   (c) Indirect material plus indirect labour plus indirect expenses
   (d) Indirect material plus direct labour plus indirect expenses

9. One of the following is a way of categorizing overhead.
   (a) Function-wise (b) Value-wise
   (c) Use-wise
10. One of the items below is not function-wise classification.
    (a) Manufacturing/production/factory overhead
    (b) Administrative overhead
    (c) Indirect materials
    (d) Distribution overhead.
11. One of the items is not considered a sales overhead.
    (a) Brokerage (b) Bad debts
    (c) Market research (d) Power
12. Are the following distribution overheads?
    - Warehouse expenses
    - Storage expenses
    - Packing and shipping expenses
    - Carriage outwards

    (a) Disagree (b) Agree

Answers

1. (c)	2. (b)	3. (d)	4. (a)	5. (c)
6. (d)	7. (b)	8. (c)	9 (a)	10. (c)
11. (d)	12. (b)			

Problems

Problem 11.1

On the basis of the following data, find out the fixed cost and variable cost applying analytical method.

Activity level (hrs)	*Budgeted overhead allowance (₹)*
6,000	16,000
7,000	18,000
12,000	28,000

Solution: Assume activity level is in "X" axis and budgetted overhead allowance is in "Y" axis.

Applying the straight line formula (Y = mX + C), where m = Variable cost per hour and fixed cost is "C". Putting the data,

$16{,}000 = m \times 6{,}000 + C$ (Equation 1)

$18{,}000 = m \times 7{,}000 + C$ (Equation 2)

Solving the both simultaneous equations, we get,

m = Variable cost per hour = ₹2.00 and

C = Fixed cost = ₹4,000.

Hence, budgeted allowance of ₹16,000 consists of fixed cost of ₹4,000 and variable cost of ₹12,000.

Problem 11.2

A company has three production departments and two service departments for a period the departmental distribution summary has the following totals:

Cost centres	*Service department*	*Production department*	*Amount (₹)*
Production		A	800
		B	700
		C	500
Service	1		234
	2		300
			2,534

The expenses of the service cost centres are calculated as given below:

	A	B	C	1	2
Service cost centre 1	20%	40%	30%	–	10%
Service cost centre 2	40%	20%	20%	20%	–

You are required to show the apportionment of overhead.

Solution:

	A	B	C	1	2
	₹	₹	₹	₹	₹
Distribution summary	800	700	500	234	300
Service centre 1	47	94	70	(234)	23
Service centre 2	129	65	65	64	(323)
Service centre 1	14	25	19	(64)	6
Service centre 2	2	2	2		(6)
	992	886	656		

Problem 11.3

The following is the budget of Superb Engineering Works for the year 1978.

Factory overhead	₹62,000
Direct labour cost	₹98,000
Direct labour hours	1,55,000
Machine hours	50,000

(a) From the above figures prepare the overhead application rates, using the following methods.

(i) Direct labour hour

(ii) Direct labour cost

(iii) Machine hour

(b) Prepare a comparative statement of cost, showing the result of application of each of the above rates to job order 555 from the under mentioned data.

Direct material cost	₹45
Direct labour: wages	₹50
Direct labour: hours	40
Machine hours	30

(ICWAI- Inter)

Solution: (a)

Sr. No.	*Items*	₹	*Direct labour hour method*		*Direct labour cost method*		*Machine hour method*	
			Formula	*Rate*	*Formula*	*Rate*	*Formula*	*Rate*
1.	Factory overhead	**62,000**	Sr. No(1)/ Sr. No. (3)	62,000/ 1,55,000	Sr. No. (1)/Sr. No. (2)	62,000/ 98,000	Sr. No. (1)/Sr. No. (4)	62,000/ 50,000
2.	Direct labour cost	**98,000**						
3.	Direct labour hours	**1,55,000**						
4.	Machine hours	**50,000**						
				0.400		**0.633**		**1.240**

(b) **Comparative statement of cost (Job 555).**

Items	*Direct labour hour method*	*Direct labour cost method*	*Machine hour method*
Direct material	45.00	45.00	45.00
Direct labour	50.00	50.00	50.00
Overhead	40.00 × 0.4 = 16.00	50.00 × 0.633 = 101.91	30.00 × 1.24 = 37.20
Total Cost	**111.00**	**196.91**	**132.2**

12 Methods of Costing

Job costing; Process costing; By-products and joint products costing; Marginal costing.

12.1 JOB COSTING

The famous statement of L.D. Miles was, "All costs are for the product". The cost has to be absorbed by the product. There are different methods of costing of the product depending upon their technicality. One of the various methods is known as job costing.

Job costing is used in different situations. They are:

1. **Mass Production:** In mass production, the product is same and manufactured on a regular basis, based on certain expected demand. If the product is not sold during the year, it is kept in inventory and will be sold in the next year. In this type, process is fixed and the materials as well. As an example, the production of car, motor cycle, etc.
2. **Batch Production:** In this type, the production is made on the basis of a selective batch. The number of items in the batch is based on certain logical basis, like economic order quantity. As an example, the toys, watch manufacturing, etc.

12.1.1 Procedure for Job Costing

The following is the procedure to be adopted for job costing.

1. On the basis of technical details, the process of manufacturing the product should be finalized. This is done with the help of planning department. Industrial engineering department is also responsible for the timing of the job.
2. The costs are collected for each job clearly identifying with production order in the following headings:

   (a) Materials

   (b) Wages

   (c) Direct expenses

   (d) Overhead

The rate for manufacturing overhead has to be finalized for different cost centres. Similarly, the administrative as well as selling and distribution overheads have to be finalized. Finally, the method and rate of absorption of overhead to the jobs should be decided on certain suitable basis.

A sample of job card sheet is shown in Table 12.1. The job sheet will indicates the items to be collected. The sheet is an example only; there can be improvement on the sheet also as per the requirement.

Table 12.1 Job cost sheet

			No:			
Customer			Work order No.			
Description			Date of Delivery			
Drawing No			Date of Commencement			
Job No			Date of completion			
Quantity			Contract Price			
		Direct Material				
Date	*Requisition No*	*Description*	*Quantity*	*Rate*		*Amount*
		Direct Labour				
Date	*Job card no*	*Particulars*				
		Production Overhead				
Details of Operation			*Hours*			
Production Cost						
Administrative Overheads						
Selling and Distribution Overheads						
Total Cost						
Profit						
Selling Price						
Prepared		Checked			Date	

The following examples will help the readers to understand the mechanics of job costing.

Example 1: A rotary shaft is passing through three departments. The detailed information is given below. You are requested to find out the selling price of the rotary shaft taking into account the profit as 20% on total cost.

Will there be any change in selling price, if profit is 20% of selling price?

Materials: ₹3500

Labour:

Department A – 50 man hours @ ₹3.00 per hour
Department B – 20 man hours @ ₹4.00 per hour
Department C – 10 man hours @ ₹6.00 per hour

Overhead:

(a) Variable overhead:
Department A – ₹4000 for 5000 man hours
Department B – ₹2000 for 2000 man hours
Department C – ₹5000 for 2500 man hours

(b) Fixed overhead: ₹12,000 for 6000 normal hours

Solution:

If profit is 20% of selling price

Selling price = Total cost + Profit

Or, Selling price = Total cost + 0.20 × Selling price

Or, 0.8 Selling price = Total cost

Or, Selling price = (1/0.8) × Total cost.

Hence, profit = (0.2/0.8) × Total cost = 0.25 Total cost

Sr. No.	*Particulars*	*Rs.*
1	**Direct Materials**	**3,500.00**
2	**Labour**	
2.1	Department A: 50 man hours × ₹3.00 = 150.00	
2.2	Department B: 20 man hours × ₹4.00 = 80.00	
2.3	Department C: 10 man hours × ₹6.00 = 60.00	290.00
	Prime Cost	**3,790.00**
3	**Overhead**	
	Variable overhead	
3.1	Department A: 50 man hours × (₹4,000/5000 man hours) = 40	
3.2	Department B: 20 man hours × (₹2,000/2000 man hours) = 20	
3.3	Department C: 10 man hours × (₹5,000/2500 man hours) = 20	80.00
3.4	Fixed overhead: 80 man hours @ ₹12,000/Normal hours 6,000	160.00
	Total Cost	**7530.00**
	Profit 20% of total cost	**1506.00**
	Selling Price	**9036.00 (A)**
	Profit 20% of Selling Price	**1883.00**
	Selling Price	**9413.00 (B)**

Example 2: A factory is operating under job costing system, make the following details in respect of the operations for the year ended December 1980.

	₹
Direct material	80,000
Direct wages	40,000
Factory overheads	50,000
Administrative overheads	34,000
Selling and distribution overheads	25,500
Profit	57,375

(a) Prepare a job sheet showing prime cost, factory cost, cost of sale and sale value.

(b) Prospect for the next year is forecast to be bright. An enquiry is being processed and it is estimated that direct materials required will be ₹ 60,000 and direct labour cost will be ₹35,000. What should be the price against this enquiry if the factory intends to earn the same rate of profit on sales? The factory recovers factory overheads as a percentage of direct wages and administration and selling and distribution overheads as a percentage of factory cost.

(Adapted from C.A. Inter)

Solution:

	₹
Direct material	80,000
Direct wages	40,000
Prime cost	1,20,000
Factory overheads	50,000
Factory cost	1,70,000
Administration overheads	34,000
Works cost	2,04,000
Selling and distribution overhead	25,500
Cost of sales	2,29,500
Profit	57,375
Sales value	**2,86,875 (a)**

Calculation of overhead rate

Factory overhead as a percentage of direct wages = 50,000/40,000 = 125%

Administration overhead as a percentage of works cost = 34,000/2,04,000 = 17%

Selling and distribution overhead as a percentage of works cost = 25,500/2,04,000 = 12%

Profit as a percentage of sales = 57,375/2, 86,875 = 20%

Profit as a percentage of cost = 57,375/2, 29,500 = 25%

Quotation against enquiry

	₹
Direct material	60,000
Direct wages	35,000
Prime cost	95,000
Factory overheads @125% of direct wages	43,750
Factory cost	1,38,750
Administration overheads @17% of factory cost	23,588
Works cost	1,62,338
Selling and distribution overhead @12% of factory cost	16,650
Cost of sales	1,78,988
Profit @20% on sales is equivalent to 25% on cost	44,747
Sales value	2,23,735 (b)

12.2 PROCESS COSTING

The manufacturing process of a product is that it goes through different processes. The output of one process is the input of the second process and it goes like that. Only at the end of the final process the product is complete. In the process costing, it is to ascertain the cost of each process. The following is the characteristic of the process costing.

1. Products should be standard.
2. Product should pass through more than one process.
3. Final product is only at the end of last process.
4. The output of one process should be the input of the second process.

The process costing is used in the following industries:

(a) Chemical industries

(b) Mining industries

(c) Manufacturing industries

(d) Public utility services.

The given examples will clarify the process costing.

Examples 3: An article passes through three processes of manufacture. From the following figures, show the cost of each of the three processes during the month of January, 1938:

Process	*No. 1*	*No. 2*	*No. 3*
Materials	₹15,000	₹5,000	₹2,000
Labour	₹8,000	₹20,000	₹6,000
Direct Expenses	₹2,000	₹7,200	₹2,500

The indirect expenses amounted to ₹8,500 and may be apportioned on the basis of wages. No account needs to be taken of stocks in hand and work-in-progress at the beginning and close of the month. The number of articles produced during the month was 240.

(Agra B.Com. 38; Bihar B.Com. 58; Vikram B.Com. 09)

Solution:

Indirect expenses to the processes are as per the wages. Hence.

Process No. 1: Process No. 2: Process No. 3: 8,000: 20,000: 6,000

Or, Process No. 1: Process No. 2: Process No. 3: 4: 10: 3

Indirect expenses are ₹8,500. So, Process No. 1 = (4/17) × 8500 = 2000

Process No. 2 = (10/17) × 8500 = 5000; Process No. 3 = (3/17) × 8500 = 1500

Process	*Total*	*Cost per article* (Output – 240 articles)
Process No. 1	₹	₹
Material used	15,000	
Labour	8,000	
Direct expenses	2,600	
Indirect expenses	2,000	
Cost of process No. 1	**27,600**	**115**
Process No. 2		
Material used	5,000	
Labour	20,000	
Direct expenses	7,200	
Indirect expenses	5,000	
Cost of process No. 2	**64,800**	**270**
Process No. 3		
Material used	2,000	
Labour	6,000	
Direct expenses	2,500	
Indirect expenses	1,500	
Cost of process No. 3	**76,800**	**320**

12.3 BY-PRODUCTS AND JOINT PRODUCT COSTING

The examples will clarify the differences among by-products, joint products and co-products.

In the coke oven batteries the mined coals are fed to get the coke for the steel industry. During the process, flue gas and the tar are available in addition of coke. The products, flue gas and tar are useful and are called by-products. Similarly, for manufacturing sugar, the fibres and molasses are produced as by-product.

When crude oil is processed, the products like motor spirit, lubricant; kerosene, asphalt, etc., have been produced. These products are called joint products.

In a automobile industry, manufacturer can produce car, bus, truck, mini-bus. Similarly, a fan industry can produce table fan, ceiling fan, pedestal fan. These are also called co-products.

Example 4: The following information has been extracted from the books of a processing company:

Joint products	*Yield in kg/tonne of coal*
Coke	585
Coal tar	72
Benzol	18
Ammonia sulphate	27
Gas	198
	900

The price of coal is ₹200 per tonne. Direct labour and overheads to the point of separation are ₹150 and ₹200 respectively per tonne of coal.

Calculate the cost of each product on the basis of weight.

(Adapted from ICWAI, Final)

Solution:

Products	*Yield in kg/tonne of coal*	*Relative weight per tonne of coal (%)*	*Costs in ₹*			
			Coal	*Labour*	*Overhead*	*Total*
Coke	585	65	130.00	97.50	130.00	357.50
Coal tar	72	8	16.00	12.00	16.00	44.00
Benzol	18	2	4.00	3.00	4.00	11.00
Ammonia sulphate	27	3	6.00	4.50	6.00	16.50
Gas	198	22	44.00	33.00	44.00	121.00
Total	900	100	200.00	150.00	200.00	550.00

12.4 MARGINAL COSTING

The objective of marginal costing is to utilize the behavioural aspects of overhead. It is not a method of ascertaining the cost of the product but is a technique to utilize the fixed and variable aspects of overhead to use as a decision making technique for the decision makers.

Marginal costing is defined as, "the ascertainment by differentiating between fixed costs and variable costs, of marginal costs, and of the effect on profit of changes in volume or type of output."

The marginal cost is defined as, "the amounts at any given volume of output by which the aggregate costs are changed if the volume of output is increased or decreased by one unit."

The marginal cost is elaborated through the following example.

	Production units: 10,000	
Particulars	*Total amount (₹)*	*Cost per unit (₹/unit)*
Direct material	20,000	2.00
Direct labour	5,000	0.50
Prime cost	25,000	2.50
Variable overhead	10,000	1.00
Marginal cost	35,000	3.500

If the sales value of 10,000 units is integrated in the marginal cost, a new picture will evolve. The difference between sales value and the marginal cost which is known as marginal contribution will indicate that how much will be available to mitigate the burden of fixed cost. If the difference is positive, it will be an indication that this will be the profit contribution of the product. If it is negative, then it incurs loss to the organization.

Using above, the picture is redrawn as below.

	Production units: 10,000	
Particulars	*Total amount (₹)*	*Cost per unit (₹/unit)*
Direct material	20,000	2.00
Direct labour	5,000	0.50
Prime cost	25,000	2.50
Variable overhead	10,000	1.00
Marginal cost	35,000	3.50
Sales value	45,000	4.50
Marginal contribution	10,000	1.00
Fixed cost	5,000	0.50
Profit (+)/Loss (–)	(+) 5.000	0.50

If the fixed cost had been ₹15,000, then there would have loss, shown as (–) ₹5000.

12.4.1 Profit/Volume Ratio

The profit/volume ratio is generally shown as P/V ratio. It is the ratio of contribution by sale volume. If it is shown as percentage, then it indicates that the so much amount of money is for profit out of 100. This is shown mathematically as below.

$$\text{P/V ratio} = \frac{\text{Contribution}}{\text{Sale value}}$$

or,

$$= \frac{\text{Sale value} - \text{Variable cost}}{\text{Sale value}}$$

The ratio can be improved in the following ways:

1. Improvement of sale volume, which can be done by:
   (a) Increase in sales price
   (b) Increase in sales volume
2. Reduction of variable cost

Sales price depends on how much market can bear. Hence, it is not possible to increase the sales price. But increase in sales volume is in the hand of engineers who can help in by making a value added product, which will increase sales. In addition, variable cost reduction is the objective of engineers.

12.4.2 Uses of Marginal Costing

The marginal costing can be used in the following situations.

1. Short-term policy like trade depression
2. In specific situations, such as:
   (i) Meeting competition
   (ii) Entry into foreign market
   (iii) Acceptance of additional volume at cut price
   (iv) Addition or deletion of product/s
   (v) Make or buy decisions
   (vi) Suspension or continuance of the business

Objective Type Questions

1. One of the industries mentioned below is not a mass production industry.
   (a) Automobile industry
   (b) Tyre and rubber products factories
   (c) Machine tool manufacturing
   (d) Cycle manufacturing industry

2. One of the industries mentioned below is not special order type industries.
   (a) Printing press
   (b) Construction equipment industry
   (c) Dies and tools manufacturing industry
   (d) TV manufacturing industry
3. In job costing the costs are collected in one of the following way:
   (a) Material, direct expenses, fixed overhead
   (b) Materials, wages, direct expenses and overheads
   (c) Wages, fixed cost, variable cost and material
   (d) Direct expenses, fixed overhead, variable overhead and material
4. One of the items is not the characteristic of process costing.
   (a) Products may not necessarily be standard.
   (b) Product should pass through more than one process.
   (c) Final product is only at the end of last process.
   (d) The output of one process should be the input of the second process.
5. Process costing is used in one of the following industries:
   (a) Machine tools manufacturing industries
   (b) Computer manufacturing industries
   (c) Automobile industries
   (d) Public utility services.
6. One of the industries follows the joint product costing.
   (a) Fan industry (b) FMCG product
   (c) Petroleum industry (d) Container industry
7. The objective of marginal costing is to utilize one of them:
   (a) The element aspects of overhead
   (b) The function aspects of overhead
   (c) The behavioural aspects of overhead
8. Marginal costing is the decision making technique for the decision makers.
   (a) True (b) False
9. P/V ratio is defined as
   (a) Variable cost by sales volume
   (b) Total cost by sales volume
   (c) Contribution by sales volume
   (d) Profit by sales volume
10. Marginal cost is used in any one of the following situation:
    (a) Normal environment (b) Entry into foreign market
    (c) New product development (d) During merger of business

Answers

1. (c)	2. (d)	3. (b)	4. (a)	5. (d)
6. (c)	7. (c)	8. (a)	9. (c)	10. (b)

Problems

Problem 12.1

A factory uses job costing. The following cost data are obtained from its books for the year ended 31st December 1974:

	₹
Direct material	90,000
Direct wages	75,000
Profit	60,900
Selling and distribution overhead	52,500
Administrative overhead	42,000
Factory overhead	45,000

(a) Prepare job cost statement indicating prime cost, works cost, production cost, cost of sales value.

(b) In 1975 the factory receives an order for a number of jobs. It is estimated that direct materials required will be ₹1,20,000 and direct labour will cost ₹75,000. What should be the price for these jobs if the factory intends to earn the same rate of profit on sales, assuming that the selling and distribution overhead has gone up by 15%? The factory recovers factory overhead as a percentage of direct wages, and administration and selling and distribution overhead as a percentage of works cost, based on cost rates prevailing in the previous year.

(C.A. Nov. 1975)

Solution: **(a)** Job cost statement for the year ending 31st December 1974.

Direct materials	90,000
Direct wages	75,000
Prime Cost	1,65,000
Factory overhead (60% of direct wages)	45,000
Works Cost	2,10,000
Administration overhead (20% of works cost)	42,000
Production Cost	2,52,000
Selling and distribution overhead (25% of works cost)	52,500
Cost of Sales	3,04,500
Profit (16-2/3% on sales or 20% on cost)	60,500
Sales Value	3,65,000

(b) Statement showing the price of the job to be taken upon 1975.

Direct material	1,20,000
Direct wages	75,000
Prime Cost	1,95,000
Factory overhead (60% of direct wages)	45,000
Works Cost	2,40,000
Administration overhead (20% of works cost)	48,000
Production Cost	2,88,000
Selling and distribution overhead (28.75% of works cost)	69,000
Cost of Sales	3,57,000
Profit (16-2/3% on sales or 20% on cost)	71,400
Price for the job	4,28,400

Note: In 1975 the selling and distribution overhead will be (25% of works cost) 1-15% = 28.75% of works cost.

Problem 12.2

A particular brand of phenyl passes through three important processes. During the week end 15th January, 1952, 600 gross of bottles are produced. The cost books show the following information:

	Process 1	*Process 2*	*Process 3*
	₹	₹	₹
Materials	4,000	2,000	1,500
Labour	3,000	2,500	2,300
Direct expenses	600	200	500
Cost of bottles	Nil	2,030	Nil
Cost of corks	Nil	Nil	325

The indirect expenses for the period were ₹1,600.

The by-products were sold for ₹240 (Process 2)

The residue was sold for ₹1,2550 (Process 3)

Prepare an account in respect of each of the three processes showing its cost and cost of production of finished product per gross of bottles.

(C.A. Nov. 1975)

Solution: **Process 1 Account**

	₹		₹
To materials	4,000.00	By transfer to process 2 A/C	8,215.38
To labour	3,000.00	(Cost per gross of bottles ₹ 13.69)	
To direct expenses	600.00		

Contd...

Contd...

To indirect expenses	615.38		
	8215.38		8,215.38

Process 2 Account

	₹		₹
To transfer from process 1 A/C	8,215.38	By sales of by-product	240.00
To materials	2,000.00	By transfer to process 3 A/C	15,218.20
To labour	2,500.00	(Cost per gross of bottles ₹ 25.36)	
To direct expenses	200.00		
To Indirect expenses	512.82		
To cost of bottles	2030.00		
	15,458.20		15,458.20

Process 3 Account

	₹		₹
To transfer from process 2 A/C	15,218.20	By sales of by-product	125.50
To materials	1,500.00	By finished product A/C	20,189.50
To labour	2,300.00	(Cost per gross of bottles ₹ 33.65)	
To direct expenses	500.00		
To indirect expenses	471.80		
To cost of corks	325.00		
	20,315.00		20,315.00

Note:

1. Indirect expenses have been charged to the three processes in the labour ratio of 30:25:23
2. Cost per gross of bottles has been calculated by dividing the cost of each process by the output, that is, 600 gross of bottles.

Problem 12.3

Calculate P/V ratio from the following particulars:

Total sales: ₹10,00,000

Variable expenses: ₹6,00,000

Net profit: 1,00,000

Solution: Contribution = Sales – Variable expenses

= ₹10,00,000 – ₹6,00,000

= ₹4,00,000

$$\text{P/V ratio} = \frac{\text{Contribution}}{\text{Sales}} \times 100 = (₹4{,}00{,}000/₹10{,}00{,}000) \times 100$$

$$= 40\%$$

13 Standard Costing

Introduction; Standard cost and standard costing; Application of standard costing; Establishment of standard cost; Standard hour, Efficiency ratio and Activity ratio; Advantages and disadvantages of standard costing.

13.1 INTRODUCTION

The actual cost of a product, process or system is the cost incurred during making those product, process or systems. It does not indicate whether it has been done efficiently or inefficiencies have crept in product, process or system. The actual cost as such is not at all useful for the management. The management is always interested to have the information that how it can be improved in future. This has been a must because of today's competitive environment. The actual does not indicate the following:

1. Proper planning for execution has been employed.
2. Materials have been efficiently used and the scrap and waste are under control.
3. Direct labour with adequate skill has been used efficiently.
4. Plant and equipment are proper and efficiently used.
5. Adequate care has been taken to keep the consumables and other overhead expenses.

In view of the above, management wants to have some measurement stick by which they will know which area to be improved and which way. This necessity brings standard cost into existence. The standard cost is not the ideal cost but cost which can be achievable.

13.2 STANDARD COST AND STANDARD COSTING

The standard cost should be developed with the help of the following:

1. Detailed technical requirement should be analysed.
2. Requirement of materials using past experience for common items and forecast requirements for new item.
3. Time and motion study for finalizing the direct labour hours.

4. Proper plant and machinery in view of the volume of production and with proper capacity utilization.

There are different types of cost. They are:

(a) **Basic cost**, which is useful for the long term.

(b) **Current cost** which is used in view of the current situation and of short term.

(c) **Expected cost** which should be used at any future situation and of short term.

The standard costing requires the following steps:

1. To develop the standard cost.
2. Ascertain the actual cost.
3. Find out the difference of actual cost with respect to standard cost.
4. Analyze the variance in different areas.

13.3 APPLICATION OF STANDARD COSTING

The standard costing can be used to various industries, from small items manufacturing industry like James Clip to airplane industry. That is to say, from ant to elephant. The standard costing is definitely a costly affair, but once it is done it will reap the harvest through benefits from day-to-day basis. The standard costing is not a one-time affair, but it has to be regularly updated because of new technology comes into being and also change of material and process. Though it looks that it is only for the big industries but small industries also can be benefited from this technique.

13.4 ESTABLISHMENT OF STANDARD COST

In order to establish the standard cost the various department of the industry should be involved. There should be close relationship and adequate involvement to make the standard cost to be as perfect as possible.

1. Design engineer
2. Production planning and control manager
3. Industrial engineer
4. Purchase manager
5. Sales manager
6. Cost accountant

Standard cost should be set for each of the following elements of cost. Those are:

1. Direct materials
2. Direct labour

3. Variable overhead
4. Fixed overhead.

13.5 STANDARD HOUR, EFFICIENCY RATIO AND ACTIVITY RATIO

The different units of measurement are being used in different industries. It may be number of pieces, kg, meter, tonnes or any other units. But in standard costing there is only one unit which is called standard hour. Standard hour is defined as number of hours to be taken to manufacture one product. If 10 units are produced in 100 hrs, then the standard hours for the product is 10.

The efficiency ratio is defined as the ratio of standard hours required for producing that quantity divided by the Actual time invested. That is,

$$\text{Efficiency ratio} = \frac{\text{Actual production converted into standard hours}}{\text{Actual hours taken}}$$

$$\text{Similarly, activity ratio} = \frac{\text{Actual production converted into standard hours}}{\text{Budgeted production converted into standard hours}}$$

The following example will make the points clear.

A manufacturing plant produces two products A and B whose details are as below. The standard hour of A and B is respectively 6 and 4.

Product	*Actual*	*Budgeted*
A	1000	1100
B	650	600

The actual hours worked were 8,150.

Product	*Actual production*	*Standard hours per unit*	*Total standard hours*
A	1000	6	6000
B	650	4	2600
		Total	8600

The efficiency ratio = 8600/8150 = 105.5%

Product	*Budgeted production*	*Standard hours per unit*	*Total budgeted standard hours*
A	1100	6	6600
B	600	4	2400
		Total	9000

The activity ratio = 8600/9000 = 95.6%

13.6 ADVANTAGES AND DISADVANTAGES OF STANDARD COSTING

Advantages	*Disadvantages*
This is a very good yardstick for measurement of performance.	If updating because of technology change, materials change or process change not done, it will indicate wrong information.
Management can pin-point areas for improvement and need not waste time.	It is a costly process and may not be used effectively by many industries.
Decision taken will be proper, timely and relevant.	

Objective Type Questions

1. The standard cost is not the ideal cost but cost which can be achievable.
   (a) Yes (b) No
2. To develop standard cost, identify the odd one.
   (a) Detailed technical requirement
   (b) Time and motion study not essential
   (c) Forecast requirements for new items
3. One of the steps is not correct for standard costing
   (a) Find out the actual cost
   (b) Assume some standard cost
   (c) Analyze the variance
4. One of the department is not required for establishing standard cost.
   (a) Design engineer
   (b) Cost accountant
   (c) Production engineer

Answers

1. (a) 2. (b) 3. (b) 4. (c)

Bibliography

1. Finance Sense – An easy guide for non-finance executives; Prasanta Chandra; Tata McGraw–Hill Publishing Company Limited: Second Print 1994.
2. Cost Accounting Problems; Raghubir Sahai Saxena; Motilal Banarsidass Delhi; Reprint, Delhi 2003.
3. Cost Accounting - Overhead Intermediate; Study Notes; Cost and Works Accountants of India; Revised 1979.
4. Cost & Management Accountancy, Method & Techniques, 1986, Arun Prasad Roy Chowdhury and Amitava Bhattacharyaya, New Central Book Agency, Calcutta; India.
5. Compendium on Value Engineering, Harold Tufty, CVS, 1983, the Indo–American Society, Kitab Mahal, 6, Raveline Street, Bombay 400001.
6. Cost Engineering, Anil Kumar Mukhopadhyaya, I.K. International Pvt Ltd.
7. Value Engineering Theory by Donald E. Parkar for the Lawrence D. Miles Value Foundation, Indian Edition, 1988.
8. Quality Circles, Concept and Techniques by B.R. Dey, Macmillan India Limited, 1988.
9. ProValue – A Productive Approach to Managing Cost by Muthai Kasi; copyright @ 2013 by Muthai Kasi; ISBN 978-1-304-21527-7.
10. Wheldon's Cost Engineering and Cost Methods: Fourteenth Edition 1978; By L.W.J. Owler and L.L. Brown; the English Language Book Society and Macdonald and Evans; London and Plymouth.
11. Technique of Value Analysis and Engineering; Second Edition; By Lawrence D. Miles; McGraw Hill Book Company, New York.

REFERENCES

1. https://blog.blackcurve.com/how-to-price-a-product-in-a-competitive-market
2. https;//en.wikipedia.org/wiki/ cost estimates# Cost_estimating_methods_and_best_practices
3. http://stats.bls.gov/ncs/home.htm, May 2010

Appendix

Present Worth Factors for an Annuity (PWA)

Year	1%	2%	3%	4%	5%	6%	7%	8%	9%	10%	11%	12%	13%	14%	15%	16%	17%	18%	19%	20%
1.	0.990	0.980	0.971	0.952	0.952	0.943	0.935	0.926	0.917	0.909	0.901	0.893	0.885	0.877	0.870	0.852	0.855	0.847	0.840	0.833
2.	1.970	1.942	1.913	1.836	1.859	1.833	1.808	1.783	1.759	1.736	1.713	1.690	1.668	1.647	1.626	1.605	1.585	1.566	1.547	1.528
3.	2.941	2.884	2.829	2.775	2.723	2.673	2.624	2.577	2.531	2.487	2.444	2.402	2.361	2.322	2.283	2.246	2.210	2.174	2.140	2.106
4.	3.902	3.803	3.717	3.630	3.546	3.465	3.387	3.312	3.240	3.170	3.102	3.037	2.974	2.914	2.855	2.798	2.743	2.690	2.639	2.589
5.	4.853	4.713	4.580	4.452	4.329	4.212	4.100	3.993	8.890	3.791	3.696	3.606	3.517	3.433	3.352	3.274	3.199	3.127	3.058	2.991
6.	5.795	5.601	5.417	5.242	5.076	4.917	4.767	4.623	4.486	4.355	4.231	4.111	3.998	3.889	3.784	3.685	3.589	3.498	3.410	3.326
7.	6.728	6.472	6.230	6.002	5.786	5.582	5.389	5.206	5.033	4.868	4.712	4.564	4.423	4.288	4.160	4.039	3.922	3.812	3.706	3.605
8.	7.652	7.326	7.020	6.733	6.463	6.210	5.971	5.747	5.535	5.335	5.146	4.958	4.799	4.639	4.487	4.344	4.207	4.078	3.954	3.837
9.	8.566	8.162	7.786	7.435	7.108	6.802	6.515	6.247	5.995	5.759	5.537	5.328	5.132	4.946	4.772	4.607	4.451	4.303	4.163	4.031
10.	9.471	8.983	8.530	8.111	7.722	7.360	7.024	6.710	6.418	6.145	6.889	5.650	5.426	5.216	5.019	4.833	4.659	4.494	4.339	4.192
11.	10.368	9.787	9.253	8.760	8.306	7.887	7.499	7.139	6.805	6.495	6.207	5.938	5.687	5.453	5.234	5.029	4.836	4.656	4.487	4.327
12.	11.255	10.575	9.954	9.335	8.863	8.384	7.943	7.536	7.161	6.814	6.492	6.194	5.918	5.660	5.421	5.197	4.988	4.793	4.611	4.439
13.	12.134	11.348	10.635	9.986	9.394	8.853	8.358	7.904	7.487	7.103	6.750	6.424	6.122	5.842	5.583	5.342	5.118	4.910	4.715	4.533
14.	13004	12.106	11.296	10.563	9.899	9.295	8.746	8.244	7.786	7.367	6.982	6.628	6.303	6.002	5.724	5.468	5.229	5.008	4.802	4.611
15.	13.865	12.849	11.938	11.118	10.380	9.712	9.108	8.560	8.061	7.606	7.191	6.811	6.462	6.142	5.847	5.575	5.324	5.092	4.876	4.675
16.	14.718	13.578	12.561	11.652	10.838	10.106	9.447	8.851	8.313	7.824	7.379	6.974	6.604	6.265	5.954	5.659	5.405	5.162	4.938	4.730
17.	15.562	14.292	13.166	12.166	11.274	10.477	9.763	9.122	8.544	8.022	7.549	7.120	6.729	6.373	6.047	5.749	5.475	5.222	4.990	4.775
18.	16.398	14.992	13.754	12.659	11.690	10.828	10.059	9.372	8.756	8.201	7.702	7.250	6.840	6.467	6.128	5.818	5.534	5.273	5.033	4.812
19.	17.226	15.679	14.324	13.134	12.085	11.158	10.335	9.604	8.950	8.365	7.839	7.366	6.938	6.550	6.198	5.877	5.585	5.316	5.070	4.843
20.	18.046	16.352	14.878	13.590	12.462	11.470	10.594	9.816	9.129	8.514	7.963	7.469	7.025	6.623	6.259	5.929	5.628	5.353	5.101	4.870

Periodic Payment Factor (PPF)

Year	6%	7%	8%	9%	10%	12%	14%	16%	18%	20%
1	1.060	1.070	1.080	1.090	1.100	1.120	1.140	1.160	1.180	1.120
2	0.545	0.553	0.561	0.569	0.0.576	0.592	0.607	0.622	0.639	0.655
3	0.374	0.381	0.388	0.395	0.402	0.416	0.431	0.445	0.460	0.475
4	0.289	0.295	0.302	0.309	0.316	0.329	0.343	0.367	0.372	0.386
5	0.237	0.244	0.250	0.257	0.264	0.277	0.291	0.305	0.320	0.334
6	0.203	0.210	0.216	0.2220	0.230	0.243	0.257	0.271	0.286	0.301
7	0.179	0.186	0.192	0.199	0.205	0.219	0.233	0.248	0.262	0.277
8	0.161	0.168	0.174	0.181	0.187	0.201	0.216	0.230	0.245	0.261
9	0.147	0.154	0.160	0.167	0.174	0.188	0.202	0.217	0.232	0.248
10	0.136	0.142	0.149	0.156	0.163	0.177	0.192	0.207	0.223	0.239
11	0.127	0.133	0.140	0.147	0.154	0.168	0.183	0.199	0.215	0.231
12	0.119	0.126	0.133	0.140	0.147	0.161	0.177	0.192	0.209	0.225
13	0.113	0.120	0.127	0.134	0.141	0.156	0.171	0.187	0.204	0.221
14	0.108	0.114	0.121	0.128	0.136	0.151	0.167	0.183	0.200	0.217
15	0.103	0.110	0.117	0.124	0.132	0.147	0.163	0.179	0.196	0.214

Compound interest factors: The periodic payment necessary to pay off a loan of Rs 1 (capital recovery) annuities (Uniform Series Payments)

Present Worth Factors for a Single Cash Flow (PW)

Year	1%	2%	3%	4%	5%	6%	7%	8%	9%	10%	11%	12%	13%	14%	15%	16%	17%	18%	19%	20%
1.	0.990	0.980	0.971	0.962	0.952	0.943	0.935	0.926	0.917	0.909	0.901	0.893	0.885	0.877	0.870	0.952	0.855	0.847	0.840	0.833
2.	0.980	0.961	0.943	0.925	0.907	0.890	0.873	0.857	0.842	0.826	0.812	0.797	0.783	0.769	0.756	0.743	0.731	0.718	0.706	0.694
3.	0.971	0.942	0.915	0.889	0.864	0.840	0.816	0.794	0.772	0.751	0.731	0.712	0.693	0.675	0.658	0.641	0.624	0.609	0.593	0.579
4.	0.961	0.924	0.888	0.855	0.823	0.792	0.763	0.735	0.706	0.683	0.659	0.636	0.613	0.592	0.572	0.552	0.534	0.516	0.499	0.482
5.	0.951	0.906	0.863	0.822	0.784	0.747	0.713	0.681	0.650	0.621	0.593	0.587	0.543	0.519	0.497	0.476	0.456	0.437	0.419	0.402
6.	0.942	0.888	0.837	0.790	0.746	0.705	0.666	0.630	0.596	0.564	0.535	0.507	0.480	0.456	0.432	0.410	0.039	0.370	0.352	0.335
7.	0.933	0.871	0.813	0.760	0.711	0.665	0.623	0.583	0.547	0.513	0.482	0.452	0.425	0.400	0.376	0.345	0.333	0.314	0.296	0.279
8.	0.923	0.853	0.789	0.731	0.677	0.627	0.582	0.540	0.502	0.467	0.434	0.404	0.376	0.351	0.327	0.306	0.285	0.266	0.249	0.233
9.	0.914	0.837	0.766	0.703	0.645	0.592	0.544	0.500	0.460	0.424	0.391	0.361	0.333	0.308	0.284	0.263	0.243	0.225	0.209	0.194
10.	0.905	0.820	0.744	0.676	0.614	0.558	0.508	0.463	0.422	0.386	0.352	0.322	0.295	0.270	0.247	0.227	0.206	0.191	0.176	0.162
11.	0.896	0.904	0.722	0.650	0.585	0.527	0.475	0.429	0.388	0.350	0.317	0.287	0.261	0.237	0.215	0.195	0.178	0.162	0.148	0.315
12.	0.887	0.789	0.701	0.625	0.557	0.497	0.444	0.397	0.356	0.319	0.286	0.257	0.231	0.208	0.187	0.168	0.152	0.187	0.124	0.112
13.	0.879	0.773	0.681	0.601	0.530	0.469	0.415	0.368	0.326	0.290	0.258	0.229	0.204	0.182	0.163	0.145	0.130	0.116	0.104	0.093
14.	0.870	0.758	0.661	0.577	0.505	0.442	0.388	0.340	0.299	0.263	0.232	0.205	0.181	0.160	0.141	0.125	0.111	0.099	0.088	0.078
15.	0.861	0.743	0.642	0.555	0.481	0.417	0.362	0.315	0.275	0.239	0.209	0.183	0.160	0.140	0.123	0.106	0.095	0.084	0.074	0.065
16.	0.853	0.728	0.623	0.534	0.458	0.394	0.339	0.292	0.252	0.218	0.188	0.163	0.141	0.123	0.107	0.093	0.081	0.071	0.062	0.054
17.	0.844	0.714	0.606	0.513	0.436	0.371	0.317	0.270	0.231	0.198	0.170	0.146	0.125	0.108	0.093	0.080	0.069	0.060	0.052	0.045
18.	0.836	0.700	0.587	0.494	0.416	0.350	0.296	0.250	0.212	0.180	0.153	0.130	0.111	0.095	0.081	0.069	0.059	0.061	0.044	0.038
19.	0.828	0.686	0.570	0.475	0.396	0.331	0.277	0.232	0.914	0.164	0.138	0.116	0.098	0.083	0.070	0.060	0.051	0.043	0.037	0.031
20.	0.820	0.673	0.554	0.456	0.377	0.312	0.258	0.215	0.718	0.149	0.124	0.104	0.087	0.073	0.061	0.051	0.043	0.037	0.031	0.026

Glossary

Accounts, code of: l) A set of numbers used to categorize major components of the whole product or components of general company overhead, 2) A set of numbers used in estimating to group costs, 3) In construction, a hierarchical set of numbers used to represent the whole project [e.g., the 16 Division CSI set or the 12 System UNIFORMAT set].

Alternative, function: A different method or way [in terms of a product] to achieve the desired result.

Analysis, function: 1) The study of product performance using two words, a verb and a noun, 2) The methodology of value analysis.

Analysis, functional: A mathematical technique originated by Vito Volterra in 1887 involving the use of integro-differential equations.

Analysis, future value: An economic technique to accumulate costs occurring at various points in time to their equivalent value at a specified future point in time.

Analysis, life cycle cost: The comparison of acceptable alternatives on the' basis of their contribution to life cycle costs using present value or annualized cost methodology.

Analysis, present value: An economic technique to discount costs occurring in future years to their equivalent current 'value and total values.

Analysis, value: 1) A method enhancing product value by improving the relationship of worth to cost through the study of function, 2) A methodology using an organized approach (job plan) with an organized effort (multidiscipline team) to provide required functions at lowest overall cost consistent with achieving required acceptance or performance, 3) The determination of the value of product functions as perceived by the user/customer in the marketplace.

Analyst, value: One who uses value analysis methodology to study a product and search for value improvement.

Assurance, value: See value analyst.

Benefit, function: The life cycle advantages, income, or revenue attributable to provision of a function.

Characteristics, essential: The minimal or necessary operational, maintenance, safety, performance, and reliability needs of the customer which must be fulfilled.

Control, value: See value analyst.

Cost, acquisition: The price paid to procure a product not produced in-house.

Cost, annualized: An economic technique to convert any defined set of present value costs to an equivalent uniform annual amount for a fixed period.

Cost, application: See customer cost.

Cost, break-even: l) The point for a given quantity of product, where the cost to purchase the product is the same as the cost to manufacture it in-house, 2) The quantity at which two competing acquisition alternatives are equal in cost.

Cost, conversion: The money expended to convert raw material, or an unfinished product, into the desired usable end product. "

Cost, customer: Product price of acquisition paid by a customer.

Cost, development: 1) The amount spent on product research, design, models, pilot production, testing, and evaluation, 2) Cost normally considered product overhead and distributed as fixed cost over an estimated number of products to be produced.

Cost, differential: The difference in the life cycle cost between two competing alternatives.

Cost, direct: 1) Cost that is directly identifiable with and attributable to the production of one specified product, 2) Cost that cannot be allocated to more than one product.

Cost, direct labour: The amount expended on salaries and wages to provide a product.

Cost, essential: All cost necessary to provide basic function.

Cost, factory: See manufacturing cost.

Costs, fixed: 1) Cost incurred which is not dependent upon the quantity of products produced, 2) Costs that do not vary with the volume of business, such as property taxes, insurance, depreciation, security, and basic water and utility fees, 3. Expenses for labour, material, equipment, and tools to produce the first product.

Cost, function: 1) The proportion of product cost allocated to functions performed by the product, 2) All costs directly associated with the performance of a particular function, 3) Costs required for the realization of a function.

Cost, general and administrative: A special classification of overhead cost normally apportioned to products "and includes salaries for executives, managers, administrative and clerical staff as well as general office supplies and equipment, marketing and as 'specified, may include advanced design, research, and other administrative costs.

Cost, incremental: 1) The difference in product cost between established incremental levels of product performance, 2) The add-on, alternative, accessory, or choice cost which takes into account the availability of existing resources when adding a new system, 3) Also referred to as variance cost.

Cost, indirect: 1) See overhead cost, 2) Also called indirect burden.

Cost, indirect labour: The amount expended on employee benefits, i.e., retirement, health insurance, vacation, other leave, unemployment compensation, and bonuses.

Cost, investment: The initial costs of product development, excluding sunk costs, which are assured to occur as a lump sum in a base year.

Cost, labour: The sum of direct labour cost and indirect labour cost.

Cost, life cycle: 1) The sum of all acquisitions, operation, maintenance, use and disposal costs for a product over a specified period of time, 2) The sum of all' costs for the development, procurenient, production, and installation of a product, as well as for its financing, taxes, operation, logistic support, maintenance, modification, repair, replacement, and disposal over the period of its useful life, 3) In manufacturing it is also referred to as the sum of development, production, and application cost, 4) The economic measure of value.

Cost, logistic support: l) The cost of spare and replacement parts and equipment with associated installation labour, 2) The cost of periodic maintenance and repair, 3) The cost for those activities necessary to plan for and provide support programmes, such as logistics, and field engineering, publications, supply support, spares, training, administration of logistic functions, and repair coordination, 4) In the military, the cost for details embracing the transport, quartering, and supply of troops.

Cost, lowest total: The lowest life cycle cost.

Cost, manufacturing: The sum of the costs expended on direct material, direct labour, and factory overhead costs for a product.

Cost, material: 1) The sum expended on raw or purchased materials needed to produce a product, 2) Normally includes the cost for packaging, inspection, shipping, and delivery of purchased materials.

Cost, non-rccurring: 1) Items of cost that represent one time expense at predicted times in the future, 2) Normally includes the cost for packaging, inspection, shipping, and delivery of purchased materials.

Cost overhead: 1) Costs apportioned to products from overhead accounts, 2) Costs which cannot be specifically and directly charged to a single product as being solely incurred by that product such as development, supervision, tooling, maintenance, heat, power, light, buildings, taxes and financing, 3) Usually fixed costs, 4) Also called indirect costs or burden.

Cost, ownership: 1) The cost to acquire, operate, maintain, repair, and dispose of the product during its period of use, 2) The cost to process the product including all finance charges, taxes, insurance, and loss of product use when it is out of service.

Cost, product: 1) The slim of manufacturing, general and administrative, and selling costs, 2) The total expenses to produce a product, 3) The transfer of money, labour, time, or other personal items to achieve an objective, 4) One component of price.

Cost, production: See manufacturing cost.

Cost, recurring: 1) Repetitive production costs that vary or occur with the quantity being produced, 2) Cost expressed in terms of a recurring direct unit cost of production of an item consisting of labour, direct burden, materials, purchased parts, expendable tooling, quality control, test, inspection, packaging, and shipping, 3) Costs which are repetitive throughout a product's useful life.

Cost, relative: 1) Differential costs between various products of functions rather than actual or absolute costs, 2) Costs which show order of magnitude only and the order of expense from greatest to least.

Cost, replacement: Future costs to replace a product or product component which are expected to occur during the product's life.

Cost, retrofit: The cost to incorporate a product improvement or necessary change into an older product.

Cost, standard: 1) Cost calculated on accepted productivity and material rates used as a norm against which to compare actual performance, 2) Costs accepted as the basis for budgeting or allocation of funds.

Cost, supplier: The price a manufacturer pays for generally off-the-shelf purchased parts, materials, and supplies as contrasted with 'subcontractor costs' who generally do some degree of product manufacturing.

Cost, total: 1) Ali cost for someone to acquire, use, enjoy, maintain, and dispose of a product, plus the time, effort and risk of buying, 2) See life cycle cost.

Cost, unneccssary: 1) Costs for functions not desired, 2) Cost for quality or performance above that needed by the user, 3) Any cost which does not contribute to value, 4) That portion of the cost of a product which does not contribute to essential functions, required performance, or marketability.

Cost, variable: Direct or indirect costs which change directly with the quantity of, or conditions under which, products are produced, as distinguished from fixed costs.

Cost, variance: See incremental cost.

Cost, vendor: See supplier cost.

Dollars, constant: Economic value expressed in terms of the purchasing power of the dollar in the base year; i.e., both inflation of cost and the time value of money are reflected through use of a discount rate.

Dollars, current: Economic value expressed in terms of actual prices each year, including inflation; i.e., present of future value of current dollars is determined by using the time value of money rather than a discount rate.

Engineer, value: One who uses value engineering methodology to study a product and search for value improvement.

Engineering, value: 1) The same as value analysis except with emphasis on application during product development and/or design, 2) The incorporation of functions on to products considered of value by the user.

Estimate, cost: 1) A product representing the art and science of predicting cost or price, 2) The summation of unit quantities of labour and material multiplied by unit costs of labour, material, overhead and profit for providing a product under a specified set of conditions.

Factor, discount: A multiplication number for converting cost and benefits occurring at different times to a common basis.

FAST, customer: The same as a technical FAST diagram except the same four supporting functions [i.e., assure dependability, assure convenience, satisfy user, attract user] always appear immediately to the right of the left scope line.

FAST, technical: 1) A diagramming technique to graphically show the logical relationships of the functions of a product, 2) Product functions displayed horizontally in diagram form using the following rules: higher order functions appear to the left answering 'why' a function occurs; lower order functions appear to the right answering 'how' a function occurs; functions occurring at the same time appear vertically below one another; scope lines indicating the scope of the value study are placed vertically; and basic function or the product is defined as being immediately to the right of the left scope line.

Function, aesthetic: 1) A function describing esteem value rather than use value, 2) A function attributable to pleasing the user rather than contributing to performance, 3) A function that indicates product features that exceed its technical utility or performance requirement, 4) Also referred to as esteem value.

Function, basic: 1) That which is essential to the performance of a user function, 2) The function describing the primary utilitarian characteristic of a product to fulfil a user requirement, 3) Also called primary or essential function.

Function, critical: A combination of the basic and selected required secondary or dependent functions defining the means used to achieve workability of the product.

Function, critical path: One of the set of basic and dependent functions that meet the 'how' and 'why' logic on a FAST diagram forming a path of essential function without which the product would not perform.

Function, dependent: 1) Lower order functions, to the right of each other on a FAST diagram, that are 'successively dependent on the one to its immediate left for its existence, 2) A function that depends on a higher order function for its existance, 3) A function which exists or is chosen in order to achieve a basic function.

Function, essential: 1) A function describing a characteristic which is absolutely necessary to a product's ability to perform the user function, 2) Also called the necessary or required function.

Function, esteem: See aesthetic function.

Function, higher order: 1) A function which is a goal rather than an objective, or an objective rather than a task, 2) A function that is more abstract than specific (i.e., 'feed people' is a higher order function than 'distribute food stamps'].

Function, independent: 1) A function that does not depend on another function or on the method selected to perform that function, 2) A function that occurs 'all the time,' [i.e., a part or assembly may have to 'resist corrosion' regardJess of what other basic or secondary function that part is performing].

Function, lower order: The opposite of a higher order function [tasks rather than objectives, specific rather than abstract).

Function, necessary: See essential function. 1) A function that indicates product features that exceed its technical utility or performance requirement, 2) Also referred to as esteem value.

Function, non-essential: See unnecessary function.

Function, primary: See basic or essential function.

Function, rcquired: See essential function.

Function, secondary: 1) The manner in which the basic function was implemented, 2) A function indicating quality, dependability, performance convenience, attractiveness', and general satisfaction beyond that needed to satisfy minimum user needs, 3) Includes supporting unwanted, unnecessary, and required functions.

Functions, required secondary: 1) A secondary function that is essential to' support the performance of the basic function, 2) A function that may result from specified design criteria.

Function, sell: 1) A function that provides primarily esteem value [such as 'improve style' or 'enhance decor'], 2) A function that may result from specified design criteria.

Function, supporting: 1) A function required by the user to make a product sell, 2) A function that increases acceptance, 3) A function to assure dependability, assure convenience, satisfy user, or attract user, 4) Also called a sell function. function, task. See user function.

Function, use: See work function.

Function, user: 1) That function performed by a product that causes its purchase by a user, 2) The function performed by an employee for the company, 3) Also referred to as a task function.

Function; unnecessary: 1) A function not contributing to the utility or desirability of the product, 2) Also referred to as a non-essential function.

Function, unwanted: 1) A negative function caused by the method used to achieve the basic function [i.e., such as heat generated from lighting which often must be cooled], 2) Also called an undesirable function.

Function, work: 1) A function which is essential to make the product or service perform as intended, 2) A function that provides use value, 3) Also called use function.

Improvement, value: The same as value analysis except with the emphasis on application to improve existing products.

Index, values: The monetary relationship of function worth to function cost [expressed as VI=FWjFC] where VI is never greater than unity, [e.g., VI=O.79], 2. Sometimes expressed as function cost [VI=FB/FC].

Investment, return on: In value analysis, the ratio of the dollars saved versus the cost of performing and implementing the study [normally expressed similar to the following: ROI=$8:I].

Life, design: The period of time intended by the designer for product use under expected levels of use, maintenance and repair, before product, disposal and/or replacement.

Life, product: The period of time product is actually used, maintained and/or repaired before being taken out of service.

Life, useful: The period of time of needed use by the customer for a/product.

Life, eonomic: l) That period of time over which an investment is considered to be the lowest cost alternative for satisfying a specific need, 2) The period of time used to justify obtaining a product from a financial standpoint as reflected by a contract period, lease, mortgage, loan agreement, statutory limitation, warranty, depreciation, method, etc.

Management, value: The same as value analysis except with emphasis on application as a management technique.

Methodology, value: The study of the relationship of cost and worth to function for any product, in a prescribed manner, using the job plan.

Mismatch, value: When function cost does not fit or match user/customer function attitude for a given function.

Model, cost: 1) A diagrammatic representation of cost based on a hierarchical structure [often work breakdown structure] of product components or functions, 2) A model which sums to the total cost of the product.

Model, energy: 1) A diagrammatic representation in a hierarchical structure of the quantity of energy consumption caused by each of the product's components or functions, 2) A model which sums to the total energy used by a product.

Model, Lee: 1) A diagrammatic representation in a hierarchical structure of the present worth or annualized expenditures relating to a product for a specified period of time, 2) A model which sums to the total life cycle cost of a product.

Model, space: 1) A diagrammatic representation, in a hierarchical structure, of the square foot (or other unit of measure) of area or space allocated to each component of a facility, 2) A model which sums to the total amount of area or space in a building.

Model, time: 1) A diagrammatic representation, in a linear form of the duration to perform all tasks associated with obtaining a product, 2) A model which sums to the total length of time to obtain a product and accounts for overlapping tirne/s of concurrent effort [i.e., a PERT chart]

Model, worth: 1) A second set of numbers, in the same units of rrreasure, superimposed on the same format used for a cost model, energy model, LCC model, space mooc, or time model which represent the worth of that model element, 2) A model which sums to the total worth of a product for the resource measured.

Objective, value: 1) The same or necessary performance or acceptance at lower cost, 2) Better performance at the same cost, or at a higher cost if a greater market share is thereby obtainable.

Owner-ship: See life cycle cost.

Period, study: The same as life span.

Plan, job: A sequential approach for conducting a value study, normally consisting of an information step and/or function analysis step review or implementation step, and an optional follow-up or measurement step.

Planning, value: The same as value analysis except with emphasis on application to strategic direction of new product development, organization strategic planning, or filling market niches.

Prevention cost: The elimination of unnecessary cost during the development stages of design or 'operations.

Price, customer: 1) The fixed sum of money or amount of service given or required to transfer ownership of products, 2) Normally the sum of product cost plus profit.

Profit, product: The difference between product price and cost where [price = cost + profit].

Progamme, value: A stated plan or procedure that uses value methodology as its basis for optimizing total cost.

Proposal, value change: 1) A change submitted by in-house personnel to improve the value of a product, 2) Also called value engineering proposal (VEP).

Proposal, value engineering: 1) The same as a value study proposal, 2) See also value change proposal.

Proposal, value engineering change (VECP): A change submitted by a contractor, pursuant to a contract provision, for the purpose of reducing the contract price or life cycle cost of the product under contract.

Proposal, value study: A recommendation, resulting from utilizing value methodology, to change a product so as to achieve greater value and/or reduce overall cost.

Quality, required: The minimal level of product performance necessary to satisfy the customer (not to be confused with the word value).

Rate, discount: The rate of interest reflecting inflation and the time value of money that is used in the discount formula to convert costs and benefits occurring at different times to a common time.

Technology, value: 1) The term used in the science or art of applying value methodology, 2) The specific method or process of value analysis for handling a cost problem.

Value, aesthetic: See esteem value.

Value, annual: Past or future costs or benefits expressed as an equivalent uniform annual amount, taking into account the time value of money.

Value, cost: Archaic, see use value.

Value, economic: 1) The relationship of benefits (utility) to cost as seen by the user, 2) The life cycle benefits as related to the cost of ownership, use, and disposal of a product, 3) Its components are use value, esteem value, and exchange value.

Value, esteem: 1) The monetary sum a user is willing to pay for functions providing prestige, appearance, and/or other non-quantifiable benefits, 2) The relative value a user places on the aesthetic functions provided by a product, 3) The monetary measure of the functions of a product which contribute to its desirability or saleability but not its required functional performance, 4) The motivated desire to possess for the sake of possession, 5) Also referred to as aesthetic value.

Value, exchange: 1) The monetary sum for which a product can be traded, 2) The market value of a product at a given point in time.

Value, function 1) The relationship of function worth to function cost, 2) See also, value index.

Value, future: The equivalent value at a specified time in the future of estimated recurring and replacement costs expected during the life of the product.

Value, good: 1) That which occurs when a product has reasonable cost and desired performance as determined by the user, 2) The lowest life cycle cost to reliably accomplish a function, 3) A relative economic comparator as .determined by the user and measured by profit and sales.

Value, market: The sale price of a product under the voluntary conditions of a willing buyer and a willing seller.

Value, maximum: The lowest life cycle cost to reliably accomplish the minimum required performance.

Value, perceived: The user's view of benefits received and the price of acquiring the product.

Value, poor: The condition which occurs when function cost exceeds function worth by a significant amount.

Value, present: The economic procedure to account for the time equivalent value of past, present, or futnre costs at the beginning of a base period.

Value, product: 1) The relationship of benefits to cost which conforms to a user's wants and resources in a given situation, 2) A specific combination of use, esteem, market, and exchange values.

Value, salvage: 1) The residual value of a product, net of disposal costs, that may derive from removal or replacement 'of the product during the study period, 2) Also the residual value from the sale of the product during or after the study period.

Value, use: 1) The monetary measure of the functional properties of a product which reliably accomplishes a user's needs, 2) The life cycle cost (worth to cost relationship) considering user function only.

Index

M

O

P

R

S

T

U

V